DANIEL HAYMOND
421-9974

WORK MEASUREMENT

GRID SERIES IN INDUSTRIAL ENGINEERING

Consulting Editor
WILLIAM MORRIS, The Ohio State University

Giffin, *Queueing: Basic Theory and Application*
Moore & Kibbey, *Manufacturing: Materials and Processes*
Morris, *Decision Analysis*
Randolph & Meeks, *Applied Linear Optimization*
Smith, *Work Measurement: A Systems Approach*

WORK MEASUREMENT
A Systems Approach

George L. Smith, Jr.

Department of Industrial and Systems Engineering
The Ohio State University

Grid Publishing, Inc., Columbus, Ohio

© COPYRIGHT 1978, GRID, INC.
4666 Indianola Avenue
Columbus, Ohio 43214

ALL RIGHTS RESERVED. No part of this publication may be reproduced, stored in a retrieval system, or transmitted, in any form or by any means, electronic, mechanical, photocopying, recording or otherwise, without prior written permission of the copyright holder.

Printed in the United States

I. S. B. N. 0-88244-136-1
Library of Congress Catalog Card Number 77-89953

3 4 5 6 2 1

To My Wife Patricia

CONTENTS

PREFACE ix

1 MEASURING WORK 1

    The Nature of Systems
    Measurement
    Errors in Measurement
    Some General Issues
    Ethical Practice
    Summary
    Review Questions and Problems
    References

2 WORK SAMPLING 15

    The Work Sampling Procedure
    Summary of the Procedure
    Sources of Error
    Computations
    Summary
    Review Questions and Problems

3 MULTIPLE REGRESSION 33

    The Multiple Regression Procedure
    Summary of the Procedure
    Sources of Error
    Computations
    Summary
    Review Questions and Problems
    Reference

4 TIME STUDY 47
The Time Study Procedure
Summary of the Procedure
Expected Performance
Sources of Error
Computations
Summary
Review Questions and Problems

5 STANDARD DATA 67
The Standard Data Procedure
Summary of the Procedure
Sources of Error
Computations
Summary
Review Questions and Problems
Reference

6 PREDETERMINED TIME SYSTEMS 83
The Predetermined Time Procedure
Summary of the Procedure
Source of Error
Computations
Summary
Review Questions and Problems
References

7 THE WORK MEASUREMENT SYSTEM. 99
Summary
Review Questions and Problems

APPENDIX

A THEORY OF WORK SAMPLING 109

B CORRELATION 111

C REGRESSION EQUATIONS 113

D MTM TABLES 115

PREFACE

This text is an intentional blend of theory and practice. It had to be a combination of the two because I cannot imagine teaching or learning theory in the absence of practice. On the other hand, if the practitioner of work measurement does not operate out of a sound theoretical base, the results can be worse than no measurement at all.

The reader is urged to approach the mastery of the specific technologies covered in the text with the attitude of a problem solver. The question should always be, "How can I best measure the performance of this man-machine system?" If the challenge of measurement colors the reader's approach to the material then the conditions will be right for maximum understanding of the material.

Lest anyone misinterpret my objective, I have chosen to exclusively address work measurement, setting aside for the time, work design and methods engineering. Of course, jobs should be engineered before measurement takes place. However, many administrators and engineers are directly involved with the problems and products of work measurement whereas few have such direct involvement with work design. Therefore, I suggest that the need for knowledge of work measurement is much more general than is the need for methods engineering.

My interest in work measurement began with Professor W. J. Richardson at Lehigh University. Wally always shared his vast experience with the art of work measurement—a task easier said than done. He also generously reviewed this entire work and his suggestions have added much to the coverage.

A significant dimension was added to my appreciation of work measurement problems by Dr. P. N. Lehoczky who, before his death, sponsored me as a fledgling arbitrator. Another source of reinforcement comes from years of continuing education and consulting with organized labor, especially the Amalgamated Meat Cutters & Butcher Workmen, The Glass Bottle Blowers Association, the International Brotherhood of Pottery and Allied Workers, and the OSU Labor Education and Research Service.

Dr. Joseph Panico, of the Maynard Training Center, was extremely helpful to me prior to the writing of Chapter 6. In addition, two quarters of Ohio State students struggled with drafts of the text and one in particular—Paul Kalmbach—reviewed each chapter with me in detail.

Finally, I appreciate the patience of Carol McDonald who typed drafts, revisions, and final copy, somehow keeping track of things I had forgotten and eliminating the inconsistencies I introduced.

1

MEASURING WORK

Ours is an age of systems. In the hindsight of history, the advent and growth of the systems view of nature might well be the attribute which is chosen to characterize this period in the progress of mankind. However, systems are not new. The universe has always exhibited certain behaviors which are more a function of the relationships among its elements than of the elements themselves. What is new, and in many cases revolutionary, is modern man's recognition of the existence of systems relationships. The growth of the systems view has already had global impact, even though formal interest in a general theory of systems only began in the mid 1940s.

This book is concerned with two levels of systems: *work systems* and *measurement systems*. Work systems consist of the integrated activities of people and machines, engaged in the production of goods and services, or in those activities which support that production. Measurement systems are somewhat more abstract. They are integrated activities of people and instruments, used to analyze, classify, and quantify certain attributes of those entities to which they are applied. More will be said about both types of systems, but basically we will be dealing with a synthesis of the two: the application of certain measurement systems for the purpose of analyzing, classifying, and quantifying work, that is, Work Measurement.

Work Measurement is an indispensable part of the planning and control of an organization. It provides information which is the basis for almost all management decision making. For example, cost accounting requires knowledge of the time needed to produce an item; production control—the time to complete an activity; queueing analysis—the time to serve a customer; and inventory control—the time to use a batch of material. In fact, without knowledge of the time needed to produce, there can be no real concept of productivity.

We begin with a brief discussion of some of the characteristics of systems which are relevant to our study of work measurement.

THE NATURE OF SYSTEMS

The late Paul Fitts (1) described a system as, "an assemblage of elements engaged in the accomplishment of some common purpose, and tied together by some common information network, the output being a function not only of the characteristics of the elements, but also their interactions." This definition particularly highlights the importance of interaction in determining system performance. It reflects the principle of *wholeness*. Hall and Fagan (2) in discussing wholeness, point out that due to interactions among elements of a system, a change in any particular part of a system, may cause a change in all the other parts of the total system. On the other hand, a condition of *independence* might exist, in which a change is not transmitted beyond the element in question. Actually, 100 percent wholeness and 100 percent independence are completely theoretical extremes of a single property. There is no sensible method of measuring the "degree of wholeness" in anything but an artificially simple system. However, the concept of wholeness is valuable in a qualitative sense. In fact, as we have seen exemplified in the definition by Fitts, many authors use wholeness to define systems.

There is substantial evidence in our language that we experience wholeness. Terms such as "mob," "the stock market," and "nation," are identified with properties of the whole system, not properties of the individual entities which are elements of the system. Work systems also possess wholeness. That characteristic will be important to the understanding of both the generalizability and the limitations of the information generated through work measurement.

Another important property of systems is *anamorphosis*. This describes the phenomenon that systems tend to increase in complexity over time. In some cases, the complexity of our societal systems has increased faster than our ability to manage them and the consequences have been severe. The experience of anamorphosis has had a profound effect on the role of managers in our society. Traditionally, the engineer has been the primary author of technology. Engineers have been, and are the men and women who apply the basic principles of science in order to design and create the "things" of our societies.

As the systems which our engineers create become more and more complex, the skills necessary to manage and to operate these systems also gain complexity. The necessity of responding to the challenge of operating, controlling, or managing complex systems has given birth to new problems in measurement. The traditional approaches to problem solving and design have been supplemented by the techniques of systems analysis and systems evaluation. Those who must know about these techniques include not only the designer but also the manager, the foreman, the union representative, and in many cases, the worker.

Systems Analysis is descriptive of the structure, the relationships, the intended function, and the possible unintended byproducts of the operation of systems. As such, systems analysis essentially deals with the internal properties of the system of interest. This analysis also includes the recognition of the hierarchical nature of systems. Thus, there is a need to define the boundary of the system, its immediate environment across that boundary, and the subsystems which are elements of every system. Most systems are superior to some systems and subordinate to others. Operationally, the realities of systems analysis require that nature be divided into three classes of elements: the elements of the system of interest, the elements which are identified with the *immediate* environment of the system of interest, and everything else. (Philosophically one might recognize that everything in the universe affects everything else, but practically, trying to deal with that concept from an engineering perspective produces functional paralysis.)

Systems Evaluation on the other hand, is descriptive of the performance of systems. The perspective is necessarily external to the system and is concerned with comparing the system's actual behavior to its intended behavior. In evaluating systems, one becomes concerned with system effectiveness—whether or not the system actually accomplishes its intended purpose, and system efficiency—whether or not the negative byproducts of the system performance offset its original intent. When we evaluate system performance, we assess its productivity in the richest sense of that term.

A class of systems which is particularly interesting because of its universality, and particularly challenging from the standpoint of measuring its performance, is man-machine systems. In our specific case we will even be more restrictive and deal with those man-machine systems which are work systems. (In order that we do not experience a communication problem, understand the traditional term "man-machine" includes both sexes.)

McCormick (3) defines a man-machine system as, "a combination of one or more human beings and one or more physical components interacting to bring about, from given inputs, some desired output." Wherever people do things, whether it be at work or leisure, awake or asleep, their ability to accomplish desired objectives can be either hampered or facilitated by interaction with machines or other inanimate objects and the methods which they employ to accomplish them. Most often, it is the quality of that interaction occurring at the man-machine interface, which ultimately determines system performance. The main problem here is that regardless of the depth of our knowledge about the machine, its operating characteristics, the way in which it reacts to various stimuli and other details of its existence, we are often incapable of anything better than the most superficial speculation regarding the level of performance which might be realized from the human element in that system. For the purpose of system analysis, performance of the total man-machine system rarely, if ever, can be inferred from a study of the machine subsystem. This means that system evaluation must be performed on the total system or at least on the man-machine subsystem.

When one first seriously attempts to evaluate the performance of a man-machine system, it is a frustrating experience. Often, this early frustration stems from the attempt to articulate "hard numbers." Unfortunately, description of man-machine systems draws, not only from the knowledge of the physical sciences, but from the knowledge of the social and behavioral sciences. The engineer can very easily develop an aversion to what might be felt to be the lack of precision manifest in the behavioral sciences. In addition, many behavioral concepts are wrapped in a shroud of confusion, mythology and elusive terminology. The analyst or practitioner is not usually looking for elaborate theories and hypothetical constructs. He or she needs specifications and operating characteristics for systems and their components. Perhaps each of us will be better equipped to deal with some of these frustrations if we relate to them in terms of the fundamental principles of measurement.

MEASUREMENT

Measurement is a natural part of the analysis and evaluation functions. Nagel (4) says, "If we inquire why we measure in physics, the answer will be that, if we do measure, and measure in certain ways, then it will be possible to establish the equations and theories which are the goal of inquiry." From our point of view, the only limitation of Nagel's statement is that it refers to physics, rather than to all scientific inquiry. However, the statement reveals that one of the inherent difficulties which we find in attempting to measure human behavior is a philosophical one. Consciously or not, analysts with technical orientations tend to reject the basic issues of measurement peculiar to behavioral inquiry and to question the rigor of the behavioral sciences. Simply stated, the analysts are comfortable with the ratio scales which characterize physical sciences. These are the scales with which many of us are most familiar. Familiarization with the variety of measurement scales which are available will add to the knowledge base necessary to appreciate the problem of work measurement.

In its purest form, measurement simply involves the orderly assignment of signs to events, objects, or properties of systems, according to rules. The act of adopting the set of rules is known as scaling. As a result, measuring almost invariably consists of either creating a scale or using an existing scale. Several examples follow.

In the scales which are most familiar, the set of signs which are most frequently used are numbers. Archer (5) has pointed out that cardinal numbers are nothing more than "an extremely convenient limitless set of distinctive signs." However, it is a common error to restrict consideration to the cases in which numbers represent quantities. When numbers do represent quantities, the scales which they represent have very special characteristics. The importance of these characteristics will become apparent in the following discussion.

The simplest type of scale is a *Nominal Scale* (Figure 1-1a). It is the simplest in that the operations needed to create a nominal scale require the least amount of information. An example of a nominal scale is the numbering of players on a football team. Fifties might be centers; sixties, guards; seventies, tackles; etc. The nonmathematical property of numbers in the nominal scale should now be intuitively obvious. (Few of us would be so naive as to propose that three centers might be replaced by two tackles, that is, $51 + 52 + 53 = 77 + 79$.) The scaling rules which apply require only the determination of equality. Equal items (defined by some stated rule) are assigned the same label. When we analyze work, one of the first steps is usually the subdivision of the total task into activity elements. In this instance we are actually creating a nominal scale.

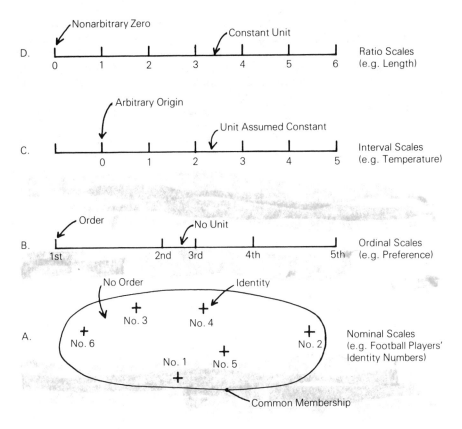

FIGURE 1-1 The Four Classes of Scales of Measurement

When and if more information becomes available, which would permit the establishment of inequalities among items, higher order scales can be generated. At this level, the numbers which had been labels, become descriptors of quantity and scales assume the characteristics of order, distance, and origin.

Order is a necessary and sufficient characteristic of a scale for normal measurement activity. An *Ordinal Scale* (Figure 1-1b) is one in which numbers are assigned to instances of a given property such that the order of the numbers corresponds to the magnitude of those instances. The act of assigning numbers is usually referred to as ranking, and positions on the scale are rank orders. Thus, if we arrange a group of trainees in order, tallest first, next tallest, second, etc., the result is a rank ordering of the height of trainees. Assigning a number one (first) to the smallest in line, two (second) to the next, etc. completes the creation of our ordinal scale of the physical trait "tallness." It is also true that reversing the ordering process and not reversing the labels would generate a scale of "shortness."

One interesting ordinal scale is the pecking order of chickens. Aggressive chicks will peck less aggressive chicks. Thus, by observing who pecks whom an ordinal scale of aggressiveness is constructed. The scale is anchored at one end by a chick who pecks all others and is never (hardly ever) pecked back, and at the other by a chick who is pecked by all others, but rarely pecks back. In the more familiar physical realm, the Mohr scale of hardness, constructed by observing which material will scratch other materials, is another example of an ordinal scale. The first step in a work measurement program might require the ranking of jobs with respect to job difficulty, thereby creating an ordinal scale. The assignment of hourly rates of pay to jobs based on their difficulty, skill requirements, or a similar attribute produces what we call an hourly pay of a *daywork* system. The level of measurement required for a daywork system to determine level of pay is an ordinal scale.

Origin is a second characteristic of scales. The only necessary constraint for this level of development of a scale is that the number zero be assigned to the zero amount of the property or to its indifference point. In our previously described scales, this would require that the numbers ascend from shortest to tallest, from least aggressive to most aggressive, and from softest to hardest. These scales are simply known as *ordinal scales with natural origins*. The most common use of the origin with an ordinal scale is in scaling properties which range from "good to bad" or "like to dislike." The point of transition from the negative to the positive degree of the property would, of course, provide a natural origin. Many authors classify ordinal scales with natural origins as a subset or ordinal scales.

The final property of interest in our discussion of measurement is distance. In the *Interval Scale*, (Figure 1-1c), order is preserved, as in the ordinal scale, but with the additional property that the size of the difference between pairs of numbers has meaning. The difference corresponds, in some way, to the magnitude of the property being scaled. This characteristic is also known as the property of equal intervals. Physical scales which exemplify the nominal scale are the temperature scales of Fahrenheit and Centigrade. It should be noted here that these scales have arbitrary zero points which do not correspond to the zero energy level.

In the behavioral sciences, raw I.Q. scores are values which possess the properties of ordinal scales. In other words, the differences in scores are an indication of inequality in intelligence (or whatever the test measures). No statements can be made, however, concerning the magnitude of the difference. The conversion of raw scores to *standardized* scores is actually an upgrading of the ordinal scale to an interval scale. In this case the interval (magnitude of the difference) in standardized I.Q. scores is a function of the proportion of the population which can be expected to attain scores within that interval.

When all three properties, *order*, *origin*, and *distance*, are present, the scale is known as a *Ratio Scale* (Figure 1-1d). Scales of time, distance, and weight are examples of ratio scales. In the realm of work measurement, many of our scaling efforts produce ratio (or at least interval) scales. The time to perform a task is an example. In order to create an incentive plan, for example, it is necessary to establish a relationship between the level of production and the amount of pay an employee or group of employees receives. Techniques for measuring work to determine that time to produce are our prime interest. Once the time is established, the knowledge can be applied to incentive pay systems, process planning and control, staffing decisions, and a host of other important management decisions.

The foregoing discussion of the nature of scales is presented for two reasons. The first is to communicate the fact that although the scales for socio-behavioral variables are in most cases nominal, ordinal, or perhaps occasionally interval scales, they are not in any way less scientific. Rather, the scales differ in their fundamental nature, and the mathematical operations which should be used to manipulate them must be chosen accordingly. To this end Tables 1-1 and 1-2 are provided, based on an analysis by Stevens (6), summarizing the characteristics of the scales and the mathematical operations which are appropriate to each type of scale.

TABLE 1-1
SCALES OF MEASUREMENT

Scale	Basic Empirical Operations	Mathematical Group Structure	Typical Examples
Nominal	Determination of equality	Permutation group $x' = f(x)$ where $f(x)$ means any one-to-one substitution	"Numbering" of football players Assignment of type or model numbers or job numbers
Ordinal	Determination of greater or less	Isotonic group $x' = f(x)$ where $f(x)$ means any increasing monotonic function	Hardness of minerals, street numbers, grades of leather, lumber, wool, etc. Daywork pay classes
Interval	Determination of the quality of intervals or of differences	Linear or affine group $x' = ax + b$ $a > 0$	Temperature (Fahrenheit or Celsius) Position Time (calendar) Performance rating scales
Ratio	Determination of the equality of ratios	Similarity group $x' = cx$ $c > 0$	Numerosity Length, density, work, time intervals, etc. Temperature (Rankin or Kelvin)

Source: From Stevens (6:25), adapted by the author.

**TABLE 1-2
STATISTICS APPROPRIATE TO MEASUREMENT ON
VARIOUS CLASSES OF SCALES**

Scales	Measures of Location	Dispersion	Association or Correlation	Significance Tests
Nominal	Mode	Information	Information transmitted, T contingency correlation	Chi-square
Ordinal	Median	Percentiles	Rank-order correlation	Sign test Run test
Interval	Arithmetic mean	Standard deviation	Product-moment correlation	t-test F-test
		Average deviation	Correlation ratio	
Ratio	Geometric mean	Percent variation	All of the above	All of the above
	Harmonic mean			

Source: From Stevens (6:27).

The second reason for the preceding discussion, is that it will become evident as we discuss the "standard" measurement techniques that we almost invariably deal with time as the basic unit of measurement. Rarely do we attempt to quantify attributes other than those comfortably expressable on ratio scales. It may be time for the engineer who faces the challenge of measuring the performance of man-machine systems to step out of the box which tradition has built.

ERRORS IN MEASUREMENT

Before discussing measurement errors, we should consider the measurement system itself. The elements of the systems which are of interest here include the instrument, the technique or method of measurement, the observer, and the object or activity being measured. The entity which we refer to as error is a product of the characteristics of each of the elements of the measurement system and of the interactions among these elements. As each technique or method is presented, specific reference will be made both to potential sources of error and ways of controlling or managing the magnitude of the error. Error cannot be eliminated; it can be managed.

The term management is not chosen lightly, because the balancing of the cost of obtaining the data against the magnitude of the error residing in the data is a management task of the first order. Much of the difficulty which has been realized in the history of measuring human work performance can be traced to a lack of appreciation of the nature of these types of error and failure to manage the measurement process.

There are two types of errors which are present in the data produced by every measurement system, errors of precision and errors of accuracy.

Errors Of Precision

Errors of precision arise from the inherent lack of repeatability in any measurement event. Even if there is *no change* in the object or activity being measured, repeated measures will vary. (The level of discrimination possible with an instrument may hide the *fact* of variability, but it does exist.) The magnitude and direction (positive or negative) of errors of precision are chance phenomena and occur randomly. In fact, given a large number of observations, the errors of precision will average to be zero, provided no error of accuracy accompanies that series of measurements. Errors of precision can be reduced to some tolerable magnitude, but they can never be eliminated. It is common to quantify errors of precision in terms of their size or magnitude, expressed as a range or an interval, and also in terms of our confidence that the true value lies within that range, expressed as a probability. For example, one might say that we estimate that a task will require 7.00 minutes to complete and we are 95 percent sure that it will actually require between 6.85 and 7.15 minutes. This means that of 100 studies, 95 would yield answers of between 6.85 and 7.15 minutes, due simply to the natural random variability in the measurement system.

Errors Of Accuracy

Errors of accuracy are caused by a bias or constant deviation of each observation from its true value. As a result of some assignable cause, the measurement system introduces an error of a fixed amount into each and every observation. Errors of accuracy are normally easy to identify and often equally easy to correct.

Referring to Figure 1-2, we can see that rifleman **B** is much more precise than is rifleman **A**. However, **B** exhibits a tendency to shoot down and to the right of center, an indication of an accuracy problem.

It is often valuable to quantify the magnitude of the errors which are inherently part of different measurement systems as an aid in determining their appropriateness for various types of applications. That quantification process is called calibration. The calibration process requires that we know in advance the magnitude of a property or amount of an activity which is to be measured. The known condition serves as a calibration standard. The process of calibration simply involves taking repeated observations or measurements of the calibration standard. The arithmetic mean of the data thus produced is compared with the standard value to provide a measure of the accuracy of the measurement system.

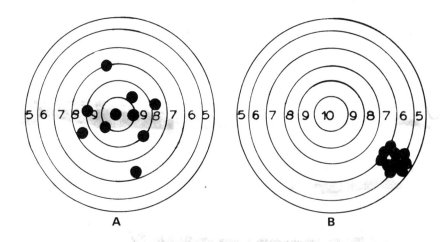

FIGURE 1-2 Errors of Precision (A) and Accuracy (B)

Measurement systems are often characterized by their inherent precision. In fact, six standard deviations is a common measure of process capability. This is the case since inaccuracy can either be eliminated or compensated for by a constant correction factor; but, as was stated earlier, inherent variability can only be held to a minimum.

In actual application, measurement systems are susceptible to both types of error. As a result, there enter errors which are above and beyond the process capability. It is helpful to know the potential sources of these errors and to use that information in choosing the measurement technique for a given application. Once the technique is selected, knowledge of errors is valuable in the management of the measurement activity in order to insure the acceptability of the end product.

SOME GENERAL ISSUES

Several attributes of human performance directly impact on the problem of measuring the performance of man-machine systems.

Adaptability And Compensatory Behavior

Students of psychology and the study of human performance are unanimous in their recognition of the extreme ranges of human adaptability and the propensity for humans to engage in compensatory behavior. Behaviorally, this means that humans will exhibit rather constant performance over fairly wide ranges of conditions. In many cases, this is due to the normal homeostatic functions of human life support systems. Thus, for short periods of time at least, humans can exhibit stable performance even when working under extreme physiologic stress. Another factor which contributes to this phenomenon is expressed through the *hypothesis of par*. That is, that human beings set arbitrary levels of performance for themselves which tend to be lower than their actual capacity. As a result, often the operator will have a reserve of spare capacity available with which to overcome factors which might be expected to depress performance, rendering the measurement function apparently insensitive to stressors on the system. This is not to imply that there is not a cost incurred due to the stressful situation. It simply recognizes that the measurement of the system performance may not capture the magnitude of that cost. Typically, if one is to overcome this dilemma, the performance of man-machine systems should be measured under loaded conditions or multiple stressors to insure that as much of the operator's spare capacity as possible is being used. Alternatively, one simply recognizes the condition of "artificially" stable performance and proceeds with the measurement. The choice, of course, depends on the purpose for measurement. In the case of establishing normal operating standards, we consciously avoid the stressful conditions because stability of performance and low physiological cost are prime objectives. If, however, we seek to establish the extremes of performance of which the system might be capable, then measurement under stress is imperative.

Trade Off Between Speed And Accuracy

In normal work, operators can slow down and decrease error rates, or speed up and increase error rates. Thus, if one measures and controls only time, the operator can hold speed constant while the error rate may fluctuate widely in response to changing conditions. In an incentive pay situation, once a standard is established an operator may speed up, apparently increasing performance, only to have a reciprocal decline in quality escape detection. The measurement techniques which are reviewed in the following chapters all assess and control on the basis of speed of performance (time). The assumption is traditionally made that quality assessment and control are independent organizational functions and are simultaneously monitored. In industrial applications where the technology has essentially been developed, this understanding was more-or-less universal and thus the issue is rarely broached in traditional discussions of methods for measuring work. However, as the need for measuring man-machine system performance expands beyond these narrow industrial beginnings, the fact of the speed-accuracy trade-off must be restated and appropriate precautions taken.

Relationship Between Method And Performance

A relationship which was alluded to earlier in this chapter, and which is one of the keys to sound work measurement, is the relationship between the method used to perform a task and the level of performance attained by the worker. The two are so inextricably related that one should not attempt to specify a time for a task without also specifying the method which gives rise to that time. The analysis and design of work is studied and practiced as an integral part of many fields of study, including motion study, work simplification, task analysis, ergonomics, bio-mechanics, human factors engineering, and value engineering, just to mention a few. Regardless of which label is attached, the basic orientation of methods analysis is—in the terms of early practitioners of industrial engineering—the search for the "one best way." The notion of *one* way has long since been dispelled, but when one seeks control of production and aspires to increased productivity, a preferred method is invariably one of the elements of the search.

Methods improvement is a rather natural human activity. Workers since time immemorial have sought alternative ways to accomplish their assigned duties, usually in an effort to find an easier, more comfortable, or more rapid means to the same end. Many work situations, however, are such that methods improvement should be conducted in an orderly or somewhat controlled fashion rather than being left to the whims of necessity and worker invention. This is not to say, however, that worker-initiated improvements are inferior. On the contrary, some excellent cooperative labor-management productivity improvement programs have worker initiation at their very foundation. The difference lies in whether the worker operates as an individual entrepreneur, looking out for him or her self, or whether job improvement is seen as part of an effort to improve total system productivity.

A formal address of method analysis and improvement technology is beyond the intended scope of this book. Our specific concern is the realization of the complementary relationship between method and performance. Much of the preparation for work measurement involves the specification of the method and the subdivision of total task into activity elements. This is a critical part of the process since it accomplishes two important objectives: (1) it enhances the precision of the measurement function, (2) it provides a record of the method which produced the measured performance. It is not unusual for a task time to be used for an extended period of time once the measurement has been approved. As often as not, situations in which the task time is judged to be incorrect are found to be a result of an unrecognized change in the method, not an inaccuracy in the original method. So placing job design aside for the moment, job analysis and specification are still an integral part of good work measurement practice.

Moving from the realm of human performance characteristics to the social dimension of man-machine system evaluation, two additional problems become important: (1) the role of organized labor, and (2) questions of ethical practice.

Union Attitude Toward Performance Measurement

There is no *union* attitude. The classic work, Gomberg's "A Trade Union Analysis of Time Study" (7) should be read by every analyst. Gomberg, writing in the late 1940s, subjected the theory and practice of time study to a critical review. This contribution was, and still is, a recognition of what were, at that time, unrecognized assumptions which were at the very foundation of that aspect of work measurement. His view is a balanced assessment of time study specifically in relation to the practice of collective bargaining.

Basically, the position of organized labor, is as an advocate for the working person. Early on, this advocacy usually manifested itself negatively and unions fought work measurement every inch of the way. More recently, however, the desirability of incentive bonuses has moved the union into a more positive posture. Most unions of any size now routinely employ industrial engineers on a full-time basis, for the purpose of insuring that the measurement of the performance of their constituents is accomplished correctly and with full consideration of the rights of the individual operator. Thus, the emphasis now can be characterized as an insistence that the measurement function be performed "by the book." This is the ultimate recognition of the fact that, if assessment and evaluation are done correctly, the operator will not be penalized. It is in everyone's best interest to see that the measurement of the performance of man-machine systems is carried out in full view of all concerned. This leads directly to the first requirement of ethical practice.

ETHICAL PRACTICE

Any time an analyst measures the performance of an operator, the operator must have been informed in advance that his or her performance was going to be measured. This does *not* mean that the analyst approaches the work place with flags waving, horns blowing, and insuring maximum disruption. It simply requires that *before* any measurement activity is undertaken, the employees affected, and, if there be any, the official representatives of those employees must be informed that a measurement program is commencing.

The corollary to the practice of notification is the issue of sharing the results of the study. Basic, good practice dictates that the results be made available to interested parties. In addition to enhancing the credibility of the results of the study and establishing a modicum of good will, the practice is consistent with the principle of providing the union with such information as might be necessary to the intelligent representation of the employees.

The U.S. Supreme Court refused to entertain an appeal of a Second Circuit Court of Appeals' unanimous ruling in support of the NLRB order which directed the company "to make available to the union certain time study data in its possession which was used in setting up standards." (Otis Elevator, 102 NLRB 72). In a second landmark ruling (J. I. Case, 118 NLRB 50) the Board stated: "An employer's obligation to grant bargaining agent's request for original time studies and job evaluation data applicable to particular job is well established . . . (Further), the Union's right to such relevant wage information is not dependent upon processing a particular grievance through the grievance procedure adopted by the parties. Nor is it dependent upon the employer's use of such information to substantiate its bargaining position as to wages." Thus, a practice which makes sense from the standpoint of positive labor relations, is supported, in principle at least, by case law. There are still managements which choose to challenge the basic issue involved. But, in your author's opinion an organization which performs the measurement function should have nothing to hide. Attempts to frustrate this orderly exchange of information on the part of management, undoubtedly would encourage the employees to speculate that perhaps the standards do not have a firm measurement base.

SUMMARY

Measurement of the performance of man-machine systems is a challenge to the analyst. Social, historical, psychological pressures, which by and large are negative, and requirements for precision which test the state of the art, combine to magnify the difficulty. Too often, work measurement has been relegated to a secondary or lower-echelon function, and we have reaped a harvest which is as good as what was sown. In the chapters which follow, specific methodologies for measuring the performance of, or establishing performance criteria for man-machine systems will be outlined. The data generated serve as the basis for many of our most crucial managerial decisions. In an age in which systems are becoming increasingly complex and the management function is becoming increasingly more difficult, we need the best performance measurement that the current technology and the art of application can provide.

REVIEW QUESTIONS AND PROBLEMS

1. Differentiate between systems analysis and systems evaluation.
2. What are the four basic types of measurement scales? Give an example of each type of scale.
3. What is the difference between errors of accuracy and errors of precision? How do these relate to the work measurement problem?
4. Why do we notify workers of impending work measurement activity?
5. Comment on the statement, "Once management finishes a work measurement study, what we do with the results is none of the union's business."
6. Of what relevance is the hypothesis of par to the problem of work measurement?
7. Explain the effect of the operator's method on the activity of measuring work.
8. Why is a systems view relevant to work measurement? How does the principle of wholeness influence the analyst's approach to measurement?
9. Explain adaptive and compensatory behavior. How do these phenomena relate to measuring man-machine system performance?

REFERENCES

(1) Fitts, P. M. (ed.), *Human Engineering for an Effective Air Navigation and Traffic Control System*. National Research Council, Washington, D.C., 1951.

(2) Hall, A. D., and R. E. Fagen, "Definition of Systems," *Modern Systems Research for the Behavioral Scientist*, W. Buckley (ed.), Aldine, Chicago, 1960.

(3) McCormick, E. J., *Human Factors in Engineering and Design*, Fourth Ed., McGraw-Hill, New York, 1976.

(4) Nagel, E., "The Causal Character of Modern Physical Theory" in *Readings in the Philosophy of Science*, Feigl and Brodbeck (eds.), Appleton-Century-Crofts, Inc., New York, 1953.

(5) Archer, L. B., *Technological Innovation—A Methodology*, Science Policy Foundation, Inforlink, Ltd., London, 1971.

(6) Stevens, S. S., "Measurement, Psychophysics, and Utility" in *Measurement—Definitions and Theories*, Churchman and Ratoosh (eds.), Wiley and Sons, New York, 1959.

(7) Gomberg, W., *A Trade Union Analysis of Time Study*, Science Research Association, Chicago, 1948.

2

WORK SAMPLING

A Work Sampling System (WSS) uses sampling techniques to estimate the proportion of time encompassed by specific elements of a given activity, based on random observations. It is the most versatile of all the measurement procedures commonly in use. Application is not limited to studies of the productivity of workers and equipment, but includes indirect and maintenance work, clerical and office operations, and supervisory and management activity. Studies are conducted over periods of from one day to several months. In fact, some studies are conducted on a continuing basis in a manner not unlike inspection for process control.

THE WORK SAMPLING PROCEDURE

Step 1. State The Objectives

The first step in preparing for a work sampling is to write a statement of the objectives. As a first step, stating objectives is hardly unique to work sampling. However, particular emphasis is given here because the versatility of work sampling may lead the user to the mistaken conclusion that a written statement of objectives is not necessary. Some examples of types of objectives include:

- To identify problem situations for further analysis
- To measure machine and manpower utilization
- To estimate unavoidable delays for time study allowances
- To measure the overall performance of a group
- To determine the severity of cyclical fluctuations
- To provide quantitative goals for first-line supervision.

Preparing the statement of objectives can be especially helpful in promoting acceptance of the results. To this end, it is important to involve first-line supervisors, other relevant managers, and in many instances, union representatives, in the process.

At the time the objective of the study is being defined, it is highly desirable to establish some quantitative measures of the output of the activity being studied with which the results of the study can be correlated (tons of material processed, holes dug, meals served, reports filed, patients treated, etc.). This serves three purposes. First, comparison of the value of the measure before the study with its value during the study will help assess whether the conduct of the study affected the performance of the system being observed. Second, the estimates for various elements can be correlated with the corresponding fluctuations in output. This can provide valuable clues to strategies for improving productivity. Third, the measure serves as a base for verifying the effect of changes instituted as a result of the study.

Step 2. Establish Activity Elements

Once the objective of the study is stated, the activity to be studied can be broken down into a set of activity elements which are consistent with the objective. What the analyst seeks is a set of activity elements which, when quantified, will provide information helpful in accomplishing the stated objective. When Tippett developed work sampling in 1934 his concern was machine idle time. In fact, the original name for the technique was "Ratio-Delay Study."

The simplest configuration of elements in such an analysis would be: (1) machine producing, (2) machine not producing. The reader can readily supply additional elements which would enrich the quality of such an analysis. Here, as in many other aspects of work sampling, the exact set of elements depends on the objective; however, there are attributes which are common to the elements of all work sampling studies. The attributes of work sampling activity elements are:

Easily Observable. Since the work sampling study is ideally an unobtrusive measurement activity, the activity elements themselves should be easily recognized by the observer. This means that, unless specific needs dictate it, the observer should be able to classify any action as one of the activity elements without interacting with the operator or the process in any way. Another benefit which accrues from well-defined elements is a reduction in the judgment required of the observer when classifying actions in one of the elemental categories. This is particularly important if the study involves the use of multiple observers or is conducted simultaneously in several different locations.

Mutually Exclusive. Any observed action must be classifiable by the observer as one and only one of the activity elements. There may be no overlapping elements.

Collectively Exhaustive. Every observed action or condition must be classifiable. Each must correspond to one of the activity elements. This means that in addition to the elements specified for the purposes of the study, two additional elements are usually included: (1) miscellaneous and (2) not observed. The miscellaneous element is self-explanatory, the not observed element is common in studies of personnel when it is either infeasible or possibly indiscreet to leave the normal work area in order to track down every last observation.

Reasonable in Number. In addition to the above attributes which are necessary for a valid study, a practical consideration should be raised. If a study is conducted, consisting of ten elements, a uniform expectation would yield a 10 percent chance of observing any activity. With non-uniform expectation, some elements would rarely, if ever, be observed. This presents a fundamental mathematical problem. The statistical approximation used to compute estimates of precision for work sampling studies requires that there be five or more observations of any activity element. If the element occurred rarely, the study which would produce five observations of that element might be erroneously judged to be too expensive to conduct, thus passing up an opportunity to quantify other, more critical, aspects of the job in question.

Step 3. Determine The Scope

The determination of the scope of a work sampling ultimately raises both cost and technical issues. Both the technical and the cost considerations are, in the final analysis, a quantitative matter; and the specific details will be covered in the computation section. Therefore, at this point the discussion will be based on the important qualitative aspects of the scope of the study. The technical issues are essentially two: representativeness and sampling error. The cost issue boils down to the total number of observations required for the study.

Representativeness. A prime consideration in the design of the study is the assurance that the sample which will be analyzed can honestly be viewed as representing the actual activity taking place. (A study performed on a Friday in a bank or a supermarket would not be representative of the *entire* work week in either institution.) To resolve the problem of representativeness, two questions must be answered to the satisfaction of all parties concerned with the results of the study: (1) was the study conducted over a period of time during which all relevant activity could reasonably have occurred? (2) Did the sampling procedure which was employed capture the essence of the activity which actually occurred? These questions should be asked *before* the study and the design should be such that the response is a comfortable, if uncertain, yes. Ultimately, the credibility of the study may rest on that *a priori* assent by the responsive parties.

Sampling Error. Any time judgments are based on a sample, error *must* exist simply because the analysis did not include every possible unit in the population being sampled. (This is not a blanket indictment of sampling since we also know that 100 percent observation is not 100 percent perfect; with sampling we simply recognize the error.) Quite simply, sampling error gets smaller as the number of observations increases. To be precise, the size of the error changes as a reciprocal function of the square root of the number of observations. Here again, the design of the study should address the cost effectiveness of an increasing sample size.

Cost of the Study. Once the fixed cost of the design of a work sampling study is incurred, the total cost increases in direct proportion to the number of observations required. Some economies can be realized in the design of the study, and these will be addressed. However, the fundamental trade-off between cost and precision and the presence of a diminishing return in precision as cost increases sets the stage for a rather pragmatic approach to determination of the scope of the study.

Four parameters are available to be manipulated in reaching a decision with respect to the scope of the study: (1) the number of observers, (2) the number of observations per tour, (3) the number of tours per day, and (4) the number of days in the study. Once each parameter is specified, the product of the four will produce the estimated number of observations in the proposed study. Given this estimated number, hypothetical computations can be made and the feasibility of the study can be determined.

Number of Observers. In most studies there will be one observer. The use of a single observer solves a basic problem of consistency in assignment of observed activity to the appropriate study element. When parallel studies at two locations are desired, or when the area in which the study takes place is large, or when the availability of a single observer is limited, then multiple observers must, of course, be used. In this event, special effort must be made to assure common understanding of activity elements. A discussion of who should be the observer follows.

Number of Observations Per Tour. In work groups with large numbers of employees it is a temptation to observe every person in the group on each tour, thus increasing the size of the sample. The problem which must be resolved is whether or not multiple observations on a single tour are, in fact, independent.

If the activities of individual workers are essentially independent, then there is considerable efficiency in taking multiple observations. However, the question of independence must be resolved in each specific instance before multiple observations are employed.

Number of Tours/Day. This parameter is, in all probability, the most misunderstood parameter in the design of a work sampling study. Somehow, the notion of sampling error is usually resolved by increasing the number of observations per day. It is not unusual, for example, to find analyst-observers dedicating virtually 100 percent of their time to a sampling study and scheduling observational tours as often as four times per hour. The result, of course, is an in-depth, "highly precise" sampling of one or two days of activity and the issue of representativeness of the study is totally ignored.

Often, the demands of other duties are such that an observer finds that even hourly tours are a burden. In this event, four, or even three or two tours per day can suffice. The observer should begin by first estimating the amount of time that a tour will require. It should be kept in mind that a tour should cover the entire physical space in which the job being studied normally occurs. In some cases, the area is confined and the observer might simply look up from his or her desk, make the observations, and continue with the task at hand. A typing pool or drafting room might present such a situation. However, a study of nursing activity in a hospital, or maintenance activity in a manufacturing plant might require a tour of several floors of a large building. Each case would dictate a different time per tour. Once the time per tour estimate is made, then one can proceed with a decision concerning the number of tours per day.

Number of Days Studied. The preceding discussion and the issues raised with respect to representativeness should aid in specifying the level of the fourth variable. Experience has shown that a minimum study ought to incorporate activity observed for a calendar month (about twenty working days) in order to capture the normally occurring cyclical activities.

The scope of the study would include both the period of time covered by the study and the total number of observations in the study. The total number of observations (n) is simply the product:

$$n = \text{\# observers} \times \text{\# observations/tour} \times \text{\# tours/day} \times \text{\# days}.$$

It is appropriate to note here that some advocates of work sampling use the technique on a continuing, even daily, basis. Operating with a reduced level of sampling and observing only a limited number of key activity elements, the results of each day's observations can be plotted over time, in the manner of a process control chart. Figure 2-1 is an example of this type of application.

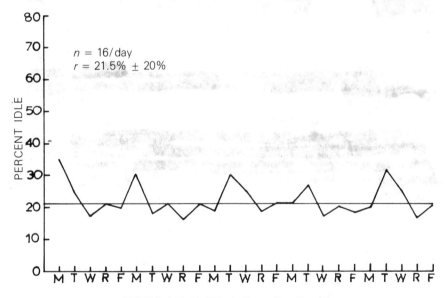

FIGURE 2-1 Daily Work Sampling Results

Having specified the four design parameters and estimated the scope of the study, the next step is scheduling tours.

Step 4. Schedule Of Observation Tours

It was pointed out that one problem which accompanies the determination of the scope of the study is that of assuring that the procedure used to obtain the sample actually captures the essence of the activity under study.

In order to meet the statistical prerequisites of work sampling, each moment within the study period must have an equal probability of being selected for observation. It has been demonstrated time and time again that human beings are incapable of spontaneously generating a schedule which is strictly random and which meets those statistical prerequisites. In addition, when schedules of observation are drawn without the impartial guide of randomization, certain periods of the day tend to be unintentionally overloaded or ignored. In order to remain objective in the study and also to meet the technical requirement that each moment have an equal probability of being selected for observation, schedules are developed with the aid of a table of random numbers. An example table is shown in Table 2-1.

There are many schemes for randomizing the schedule of observations. The one explained here was selected for its simplicity and its flexibility. It starts with the same information used to estimate the total number of observations, an estimate of the number of tours per day. Two questions must be answered. First, how long will it take to complete a single tour of the area being observed? This estimate is used to insure that the time interval between successive tours is never less than the time required to complete the first tour. For observers from outside the area or for large areas, a variety of tour routes must be used and sufficient time allotted for their completion. For local observers in a small area, a scan may accomplish a tour with little or no physical movement by the observer. The second question is, "How many tours per day can the observer reasonably be expected to conduct?" One or two per hour are usually a maximum; two or three per day a minimum.

Once these two parameters are specified, the random number table comes into use. Enter the table in any row or column. Continue along the file until the end is reached. Move on to the next, until a sufficiently large sample of usable numbers is extracted. An example will complete the explanation.

Assume a tour is expected to last fifteen minutes and that the observer can handle an average of one tour per hour. The work day is 8:00 to 5:00 with lunch from 12:00 to 1:00, and there are no scheduled breaks. Our process is one of randomizing within the time interval. For hourly intervals, random numbers from 00 to 59 (60 minutes/hour) would be used. Consider the following set of two-digit numbers: 42, 37, 77, 99, 90, 89, 85, 28, 63, 09, 09, 10, 51, 02, 52. The numbers 77, 99, 90, 89, 85, and 63, are eliminated at the outset. The schedule which would then be developed is shown in Figure 2-2. However the tours at 3:51 and 4:02 are within the fifteen-minute tour duration, so the next number is selected. Since 4:52 is within eight minutes of the end of the day, the tour would continue for the first seven minutes of the next day.

Using the principle of randomization within the interval, any schedule can be accommodated. Two tours an hour would randomize within thirty-minute intervals using digits from 00 to 29. A schedule of tours every two

WORK SAMPLING   21

TABLE 2-1
RANDOM SAMPLING NUMBERS

	A	B	C	D	E	F	G	H	I	J	K	L	M	N	O	P	Q	R	S	T	U	V	W	X	Y
1	90	78	82	54	47	20	83	80	10	41	35	22	23	03	98	79	74	41	35	05	78	73	95	47	83
2	78	58	68	87	41	11	08	81	29	89	71	23	10	01	79	25	06	00	45	80	64	70	95	34	29
3	51	42	21	03	88	20	05	35	93	00	68	12	09	55	09	36	54	95	22	82	48	30	09	56	87
4	93	15	07	60	86	67	37	94	24	35	82	44	19	92	96	21	84	29	04	29	83	32	05	10	48
5	27	12	31	66	62	09	54	17	31	23	27	30	37	36	79	75	50	39	57	12	67	23	22	09	33
6	79	44	83	55	47	96	50	93	56	82	58	16	35	18	87	64	08	22	47	93	86	43	43	30	17
7	89	73	43	91	03	57	91	35	40	64	13	61	94	37	16	09	93	96	25	87	30	23	42	54	31
8	29	30	90	00	58	15	99	93	33	67	80	08	59	21	66	13	54	56	85	25	05	32	03	52	52
9	97	33	17	26	25	04	73	18	10	05	34	40	32	65	07	28	68	29	31	97	89	57	95	55	16
10	07	15	44	92	47	28	50	93	03	53	37	70	19	68	59	95	39	87	90	46	98	64	46	24	71
11	82	50	35	50	80	23	67	81	25	02	83	08	12	70	00	25	31	33	80	06	19	86	14	59	27
12	59	21	86	16	30	27	85	16	26	34	50	15	87	22	69	71	36	95	90	76	90	99	79	63	21
13	04	19	60	33	05	29	02	33	74	56	38	84	21	07	35	93	54	70	18	47	14	62	75	45	02
14	96	91	44	09	94	06	89	50	88	83	82	50	11	82	51	30	68	91	06	28	86	65	17	45	20
15	31	71	03	53	38	94	02	52	72	15	44	49	53	42	43	00	36	97	67	64	12	27	46	00	18
16	03	70	22	67	59	98	10	64	68	08	79	06	89	48	41	85	72	10	87	24	96	04	20	68	00
17	08	45	79	46	89	74	73	67	60	15	70	37	61	44	07	67	89	81	54	26	57	17	63	27	74
18	37	80	05	75	64	48	51	68	68	27	71	75	45	32	27	76	35	26	58	88	67	74	48	90	94
19	90	63	56	69	37	19	74	48	63	31	52	36	84	40	66	72	66	03	41	87	65	29	12	36	64
20	22	69	38	02	88	89	71	43	01	87	41	79	42	99	29	41	08	47	32	19	45	29	59	69	90
21	05	79	69	67	64	36	14	82	65	26	40	51	63	42	48	85	48	34	12	04	33	26	52	26	52
22	48	91	53	03	82	64	24	06	31	03	97	44	82	24	89	88	48	66	54	10	41	27	09	11	61
23	94	64	97	27	25	62	23	94	40	54	56	32	97	78	90	58	86	41	75	19	42	90	85	36	68
24	15	85	82	52	08	52	96	26	92	88	93	11	03	23	52	78	23	57	85	43	53	90	42	22	22
25	09	81	37	66	56	99	08	59	19	48	29	69	21	64	95	12	08	15	24	45	59	25	22	76	96
26	43	83	99	02	76	12	16	45	52	66	35	70	93	09	52	75	40	34	35	62	65	42	27	20	59
27	31	98	09	80	62	75	26	64	57	26	46	41	47	90	97	99	46	10	51	42	73	28	98	89	91
28	81	35	42	62	84	37	02	59	78	16	17	96	05	71	39	88	05	34	05	92	22	43	89	66	89
29	97	95	56	39	75	65	47	61	86	33	14	88	55	33	69	70	87	79	94	46	17	61	72	27	01
30	37	63	35	93	23	17	30	14	51	51	17	28	21	74	67	12	11	57	19	27	38	70	73	82	92
31	39	22	96	00	48	52	49	62	09	40	08	30	27	54	70	96	06	52	12	80	36	12	38	68	05
32	61	29	84	34	51	60	19	77	82	16	64	45	02	27	04	65	55	90	95	04	20	39	29	96	28
33	38	84	18	10	29	19	09	66	06	78	37	09	60	50	21	52	72	01	52	70	29	65	05	37	16
34	64	29	48	04	08	55	72	25	25	77	54	26	27	24	39	66	67	06	40	00	99	35	70	69	58
35	64	02	32	99	63	62	42	89	32	20	81	14	08	40	45	22	15	37	49	38	96	51	19	08	27
36	13	83	39	51	30	31	49	94	83	66	02	50	95	18	98	58	84	90	58	81	00	40	91	12	46
37	83	30	90	09	35	41	12	87	93	66	85	96	20	65	34	23	13	05	41	01	91	48	95	59	45
38	46	63	53	97	63	18	86	37	56	20	35	62	66	11	37	30	91	89	97	51	64	78	06	95	65
39	54	43	40	02	41	55	70	52	96	87	02	82	61	21	88	60	65	98	42	09	03	61	20	83	01
40	27	18	65	62	01	97	45	79	51	37	74	47	20	11	48	97	93	73	86	50	46	61	95	01	24
41	45	42	16	13	20	34	51	08	71	52	39	17	71	39	84	97	27	72	49	42	81	62	32	87	22
42	35	92	97	02	34	93	32	95	81	13	92	05	40	70	95	71	66	61	24	08	77	32	73	66	79
43	60	55	35	57	24	52	95	84	90	64	38	39	72	70	17	98	42	85	96	67	41	11	83	17	78
44	43	17	21	09	60	58	86	12	31	11	66	61	43	96	00	93	97	00	15	20	37	96	73	56	63
45	07	85	74	58	28	38	74	68	32	61	87	14	71	83	47	90	11	96	70	08	67	04	34	46	08
46	33	00	29	08	87	42	59	40	24	97	44	99	13	56	87	95	02	47	97	89	23	51	45	37	83
47	97	14	00	42	23	72	03	19	02	41	11	23	36	98	32	19	91	42	03	58	62	23	74	45	06
48	68	58	32	80	82	40	49	71	83	37	93	49	99	60	72	88	14	26	88	95	48	69	35	40	63
49	39	87	38	16	06	82	92	62	32	75	67	64	50	49	39	29	55	53	92	97	04	48	60	53	90
50	37	73	01	84	87	42	88	30	93	75	01	18	34	73	30	28	44	28	18	01	00	38	26	38	57

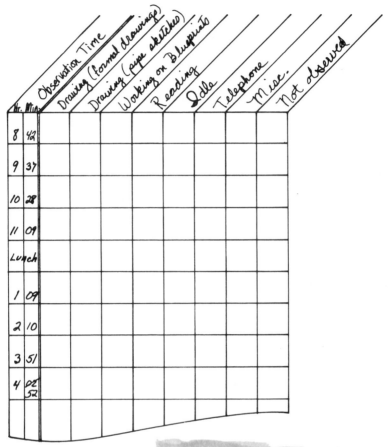

FIGURE 2-2 Work Sampling Data Form

hours requires three digit numbers 000 to 119 (120 minutes). If three-digit tables are not available, use the first digit from the next column. An example of two-hour intervals appears in the case which follows in the computations section.

Step 5. Design The Forms

The recording of observations and the accuracy of the transcription of results can be greatly enhanced if attention is paid to the design of data collection forms. An example of a form appears in Figure 2-2. Also, consideration should be given to mark sensing, optical character recognition, or some other form of machine processing to reduce error and speed the availability of results.

Step 6. Notify The Appropriate Parties

The issues associated with this step were discussed in Chapter 1. This step may not be omitted. The negative consequences far outweigh any benefits which might be imagined to accrue. Figure 2-3 shows an example of a memo announcing a work sampling study.

MEMORANDUM

From: A. T. Jones, Plant Manager
Subject: Notice to all Warehouse Employees
Date: February 10, 1975

A study of warehousing operations is being conducted in conjunction with plans to enlarge the warehouse. Part of this study includes a work sampling of the activity of inventory clerks, to determine staffing requirements in the expanded facility. The study will begin in the next few weeks and is expected to take approximately one month.

We appreciate your cooperation in this effort. If there are any questions, contact your foreman.

FIGURE 2-3

Step 7. Conduct The Study

Once the preliminary steps are completed, the actual conduct of the study is simple and straightforward. For this reason, it is not necessary that the observer be experienced in work measurement. The only necessary characteristic is that the observer know the task involved, be able to classify observed activity on sight, and be unbiased in the classification process.

Given this characterization, it should be clear that the analyst who designs and supervises the study need not serve as the observer. Common practice generally places the industrial engineer or engineering technician in the observer-analyst role. This practice is acceptable if the analyst normally spends a large portion of his working day in the area where the study is conducted for purposes other than those of the work sampling study. However, if the only reason for the observer's presence is the study, the unobtrusive quality of the study will probably be destroyed. Workers, being aware of the study's existence, will naturally want the results to reflect well on themselves. Therefore, the presence of an observer may substantially alter the activities to be observed. This problem is, of course, minimized if the observer is normally present, and the observations are then taken at random times. For this reason, there is substantial justification for preferring an observer who is frequently in the area of the study over someone who simply has organizational responsibility for conducting the study, or the technical expertise to analyze the results. In this author's opinion, the single most appropriate observer is the supervisor who has immediate responsibility for the work unit being studied. (This discussion does not apply to working foremen or group leaders.)

The use of the supervisor as an observer should not be rejected arbitrarily. The supervisor should have been included in the planning of the study from the start, whether or not he or she serves as the observer. Therefore, knowledge of purpose is assured. The supervisor's presence in the immediate area of the study is, hopefully, not an unusual occurrence. Finally, the supervisor will undoubtedly be a key person in implementation of the results. All things considered, the supervisor is the prime candidate for the observer role.

Unfortunately, in many organizations the first-line supervisor has allegiances which are closer to the rank and file worker than they are to management, making him or her of suspect value as an unbiased observer. The problem arises quite understandably. Many organizations promote from the rank-and-file into first line supervision. However, whether supervisors are up from the ranks or not, they are totally dependent on the cooperation of their subordinates for the single, most observable index of their supervisory performance—production. This spawns the mixed allegiance. Such a situation is unfortunate, because, if top management is suspicious of the allegiances of first line supervisors, there are probably other suspicions, all of which militate against a cooperative approach to work measurement. In most situations, however, a convincing case can be made for the use of the supervisor as an observer. The key is that the supervisor believes that the measures taken as a result of the study will be beneficial to the work unit.

Suppose an overly zealous supervisor/observer overestimates the proportion of observations classified as productive activities. On the basis of the study, management would, in all likelihood, expect output which is actually beyond the capacity of the group. (An expectation which could place a serious hardship on both the group and the supervisor.) If, on the other hand, the observer underestimates the productivity, steps which might be instituted to raise productivity could place an equally severe burden on the group. It is, in the long run, in the best interest of the work group as a whole, and the supervisor in particular, that the study show exactly the conditions which actually exist. This attitude must be communicated to both the observer and those who are being observed.

There is one additional problem which will undoubtedly be raised by the first-line supervisor when it is proposed that he or she serve as the observer. Most supervisors will protest that they are over-worked, and that it is an imposition to add observer duties to the existing work load. This argument is paradoxical. The observer has two duties: that is to observe the activity in the work unit at predefined random intervals, and to record the observations on the prepared form. If a supervisor's collateral duties are so demanding that he or she cannot take time to observe work group activity once or twice an hour, the role of supervision in the organization ought to be reconsidered before a work sampling study is undertaken. In fact, one might question the appropriateness of the title "supervisor" for a individual whose work precludes him or her from touring and observing the activity of those being "supervised" at least two or three times per day.

It must not be ignored, however, that random observations, if performed on time, can be disruptive. To this extent, the study does place an additional burden on the observer. Many potential observers are concerned about this disruptive quality. It should be noted that the arrival of a scheduled tour time is not an excuse for rude behavior (hang up phone, interrupt conversation, etc.). The reason for randomization has been explained. In operationalizing that objective, the observer need not be a slave to the precise schedule times; however, the schedule may not be revised to conform to the whims of the observer's work schedule.

Once the tour begins, the observer simply makes visual contact with the individuals or equipment being studied. At that time an instantaneous classification of the activity should take place and the appropriate marking placed on the data sheet. If, for some reason, the observer should want to interact with the operator, that interaction should only take place after the observation is recorded. The tour continues until the area to be surveyed has been traversed. At that time, such "not observed" entries as may be required are made and the observer may return to his or her duties until time for the next tour.

Tours should not normally be made during scheduled breaks. If a break occurs during a tour, the tour should simply be discontinued until after the break, at which time the remainder would be completed. For studies in which a "tour" involves no need for physical movement by the observer, any random times which concur with a break should simply be ignored.

SUMMARY OF THE PROCEDURE

A work sampling study, which estimates the proportion of time spent in given elements of activity, requires five items of information: (1) a stated purpose for the study, (2) some specific activity elements which describe the task to be studied, (3) an assessment of the scope of the study, (4) a randomized schedule of observational tours, and (5) some well-defined measures of output with which the sampling results can be correlated.

SOURCES OF ERROR

There are basically three types of error which can affect the results of a work sampling study. Each type of error has been referred to in earlier discussions. They are mentioned here to highlight the measures which are available to minimize their effects.

Sampling Error

Sampling error is present in all sampling studies. The amount of the sampling error is measured by the size of the confidence range. This means that the strategies which are available include: increasing the number of observations in the study, increasing the size of the confidence range, or relaxing the confidence level. The trade-offs which are involved are obvious and the analyst is cautioned to keep in mind the purpose of the study in establishing any of the decision parameters which relate to sampling error control.

Bias

Bias in work sampling studies has two sources: the observer and the individual who is being observed. The act of classifying observed activity into the appropriate activity element requires the exercise of judgment on the part of the observer. As a result, the opportunity is always present for predisposition toward the desired outcome of the study to introduce bias. Observer bias can be controlled by gaining the cooperation of the observer in advance of the study, and by precise definition of the activity elements. In the event that multiple observers are used, group training is recommended. The training is best accomplished by having all potential observers participate in a series of trial tours as a group. Each observation is classified independently by the observers. Following this, the group shares their judgments and any differences are discussed and resolved. With immediate feedback of this type, uniform judgments can be achieved rather rapidly.

Operator bias is more difficult to control than is observer bias. The principal strategy relies on the randomization of the times of the tours and the actual tour routes to minimize the ability of operators to anticipate the arrival of the observer and modify their behavior. This should be supplemented by a concerted effort by management to instill confidence among the workers that their interests will be protected in any action which might result from the study. The analyst should also provide for an ongoing monitoring of the quantity of production before, during, and after the study, whenever possible. Not only will the information be helpful in the determination of potential changes, but it will provide a measure of the degree of impact the study had on the output—a good indication of possible operator bias.

Representativeness

This third type of error is possibly the most insidious of all errors. Since the error is not one of obvious bias or imprecision, it is possible for lack of representativeness to go undetected. In fact, as was pointed out earlier, it may happen that the desire to expedite the study leads the analyst to increase the daily sampling rate and shorten the length of the study. From the standpoint of producing a study with results that can be generalized, such a strategy is extremely undesirable. In fact the result can be seriously misleading. In controlling for errors of representativeness, there is no substitute for intimate knowledge of the work being studied and a deliberate and systematic approach to the measurement task at hand.

COMPUTATIONS

After the data are collected, the next step in the analysis is charting and the computation of the relevant statistics. Each activity element can be represented by a point estimate of the proportion of observations assigned to that element and a range of values around that estimate expressing the error of precision. (In the case of sampling studies, the inherent error of precision due to sampling is simply referred to as sampling error.)

The estimate of the proportion of time is simply the number of observations within the element, divided by the total number of observations in the study. The range of sampling error is based on the application of the binomial distribution to each activity element, and computations use the normal approximation to the binomial. (For a more complete development see Appendix A.) If p = the proportion of observations which were classified in a given activity element, and n = the total number of observations taken in the study, then the range of the sampling error, r, is:

$$r = p \pm k \frac{\sqrt{p(1-p)}}{\sqrt{n}}$$

where k is the confidence factor which depends on the desired level of confidence alluded to in Chapter 1.

$k = 1.64$ is associated with "low" confidence (90%).
$k = 1.96$ is associated with "moderate" confidence (95%).
$k = 3.00$ is associated with "high" confidence (99.72%).

To meet the statistical requirements for computing the range, there must be five or more observations of the element. If there are less than five, the analyst may: (1) search for another related element and combine the two into one meaningful element, (2) take more observations until there are five or more, or (3) simply not compute a confidence range.

In the early development of Shewharts' process control charts, a conventional practice regarding confidence factors developed. Essentially, it was suggested that 95 percent confidence ($k = 1.96$) balanced the cost of concluding a process was not in control, when in fact it was, with the cost of not detecting loss of control, when it had actually occurred. The rationale is not particularly relevant to work sampling (even though one could develop a cost model and select a factor to balance cost). However, the mystique of the 95 percent confidence level continues, and many an analyst finds $k = 1.96$ being preordained by a well-intentioned manager. If that is the case, there are a sufficient number of other ways to reduce the required sample size. The issue of confidence level need not be a matter of contention.

The following example is based on a study of inventory clerks in a large warehouse. The purpose of the study was stated in the memorandum (Figure 2-3) released to the warehouse several weeks before the start of the study.

In determining the scope of the study, the foreman suggested that his responsibility for shipping and receiving in addition to the warehouse limited him to four tours per day. He also estimated that a warehouse tour would take about fifteen minutes. Since there were three clerks picking and stacking in the warehouse on one shift, a twenty-day study had a scope of:

$$\text{observers} \times \frac{\text{observations}}{\text{tour}} \times \frac{\text{tours}}{\text{day}} \times \frac{\text{number of}}{\text{days}} = \text{Scope}$$
$$1 \times 3 \times 4 \times 20 = 240$$

The activity elements which were relevant to the decisions for warehouse expansion could have been limited to productive and nonproductive activity and not observed. In order to take advantage of the fact that the study was being conducted, however, ten elements were identified. They are shown in Figure 2-4.

There was a general feeling among management that due to lack of automation, among other things, the two clerks spent at least one-third of their time walking, and thus were severely overloaded. An estimate of the study outcome was made to assess the confidence range if, in fact, walking occurred 33 percent of the time. A confidence level of 95 percent was specified.

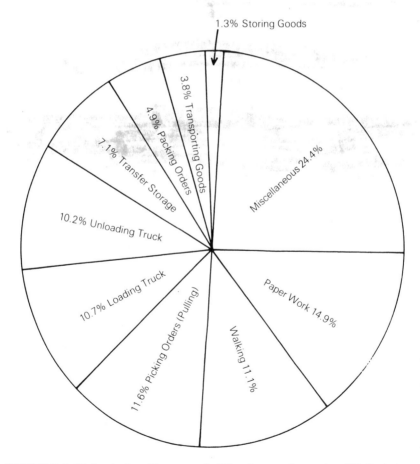

FIGURE 2-4 Major Labor Elements in Warehouse Operation (% Estimates Obtained by Work Sampling)

$$r = .33 \pm \frac{1.96\sqrt{(.33)(.67)}}{\sqrt{240}} = .33 \pm .059$$

range: .271 - .389 or 27.1% to 38.9%

It was decided at that time to proceed with the study as planned, with the idea that a second month of data might have to be collected to reduce the sampling error to about ±3 percent (a value thought to be more acceptable to top management). The randomization required three-digit random numbers. In generating the tour times shown in Figure 2-4, columns **F** and **G** of the random number table (Table 2-1) were used. Proceeding down the columns using 000-119, the following usable numbers were generated: 110, 095, 047, 068. For a 7:30 to 4:00 operation, the tour times were:

$$
\begin{aligned}
7{:}30 + 110 &= 9{:}20 \\
9{:}30 + 095 &= 11{:}05 \\
\text{Lunch} \quad (11{:}30 \text{ to } 12{:}00) & \\
12{:}00 + 047 &= 12{:}47 \\
2{:}00 + 068 &= 3{:}08
\end{aligned}
$$

The results of the study, based on sixty observations per week were: 18 percent in week **1**, 10 percent in week **2**, 13 percent in week **3**, and 7 percent in week **4**.

For the week, then:

$$p = \frac{29}{240} = 12.1\%, \quad r = .121 \pm 1.96\frac{\sqrt{(.121)(.879)}}{\sqrt{240}}$$
$$= .121 \pm .041 \text{ or from } 8\% - 16.2\%.$$

Thus, the true percent time spent walking is between 8.0 percent and 16.2 percent, indicating that the estimate of 33 percent was highly inflated. With respect to the confidence range, an estimate was needed to determine the number of observations required to reduce the range from ±4.1 percent to ±3 percent. Assuming that the estimate of 12.1 percent will hold, then, it is desired that:

$$1.96\frac{\sqrt{(.121)(.879)}}{\sqrt{n}} = .03, \quad \sqrt{n} = \frac{.639}{.03}, \quad n = 454$$

(Notice that the numerator remains unchanged in this computation. For computational simplicity, the numerator and the denominator have been separated in all formula statements.) At the rate of sixty observations per week, about three and one half weeks more study would be required to obtain a ±1 percent reduction in the confidence range, with what might objectively be considered to be no practical increase in the information management would possess.

Notice the weekly estimates of p. It might be suggested that there is a trend during the month (18, 10, 13, 7 percent). To see whether or not the

18 or the 7 percent are representative of the basic activity estimate of 12.1 percent, a third computation is needed, based on a sample size, $n = 60$, for weekly observations.

$$r = .121 \pm \frac{.639}{\sqrt{60}} = .121 \pm .083, \quad r = 3.9\% \text{ to } 21.1\%$$

Therefore, more information would be necessary to establish whether or not the amount of walking decreases toward the end of the month.

One additional modification in the decision parameter is available in order to reduce the confidence range: that is a relaxation of the confidence level. Recall the discussion concerning the choice of confidence level. The confidence factor used in the preceding computations was "moderate" ($k = 1.96$), that is the 95 percent confidence level. The range for "low" or 90 percent confidence is:

$$r = .121 \pm 1.64 \frac{\sqrt{(.121)(.879)}}{\sqrt{240}} = .121 \pm .035, \text{ or}$$
$$r = 8.6\% - 15.6\%.$$

By simply reducing the confidence level from 95 percent to 90 percent, the confidence range shrank from ± 4.1 percent to ± 3.5 percent. This is a modest increase in the stated level of precision, and it was accomplished by relaxing a specification which might well have been established by convention.

This act of systematically varying the parameters which are under the control of the analyst for the purpose of assessing their effect on the cost of the study is called a sensitivity analysis.

In the process of making the computations which are necessary to analyze the results of a work sampling study (or any other study, for that matter), the analyst should be alert to performing the type of sensitivity analysis which has been illustrated in the foregoing discussion. Very often decision makers apply conventions to prescribe the levels of decision parameters. A sensitivity analysis can be invaluable in communicating the cost which might be incurred by arbitrarily specified conventions.

SUMMARY

Work sampling is the most macroscopic and the most versatile of the measurement techniques. Due to its unobtrusive nature, it is a candidate for surreptitious work measurement. Its use in this manner could destroy the wide acceptance it currently enjoys with labor. On the other hand, the full potential of work sampling for control of indirect labor cost has yet to be tapped to any marked degree. A few analysts, trained in the fundamentals of the methodology can manage a program which utilizes a large number of observers to achieve comprehensive coverage. The reader is encouraged to explore new and interesting applications and take advantage of all that the technique has to offer.

REVIEW QUESTIONS AND PROBLEMS

1. Prepare a graph of the relationship between p (the proportion of occurrence) and n (the number of observations) for a work sampling study. Use a "low" confidence level and a 10 percent range of sampling error (± 5 percent).
2. Plot, on the same graph, $n = f(p)$ for a 4 percent range (± 2 percent).
3. Comment on the implications of the above graph for the practice of work sampling.
4. A study of three hundred observations in a bank yielded fifty observations of tellers standing idle. Compute the range of sampling error for all three levels of confidence.
5. Using column **p** in the random number table in your text, and starting at the top, prepare a schedule for a work sampling study averaging one observation per half hour. Prepare a one day schedule (sixteen tours). Tours only take five minutes. The day starts at 7:00, ends at 4:00 and lunch is from 11:00 to 12:00.
6. Estimate the scope of the study in problem five if two observers are used, we can obtain five observations per tour, and have twenty days to conduct the study.
7. If an activity element in the study outlined in five and six occurred 20 percent of the time, are there sufficient observations to be 95 percent sure that the true occurrence is between 18 percent and 22 percent?
8. A work sampling study based on 144 observations estimated that the admissions desk in the hospital is occupied 80 percent of the time. What is the "low" confidence range? How many observations would be required to halve (½) the size of the range?
9. What are the characteristics of activity elements for a work sampling study?
10. Why are tour times randomized?
11. What are the major sources of error in work sampling? How can we reduce their impact?
12. Discuss the pros and cons of having first line supervisors conduct work sampling studies.
13. To reduce the cost of a study, would you rather reduce the confidence level or increase the range? Does it make any real difference? Any psychological difference?

3

MULTIPLE REGRESSION

A Multiple Regression System (MRS) applies the statistical technique of multiple regression, to analyze the output of one or more operators engaged in a variety of tasks, in order to determine the time used to produce a single unit of output or production for each of the tasks, based on the assumption that the total time spent is a linear combination of the time spent on each task and that the coefficients of multiple regression adequately represent those times.

THE MULTIPLE REGRESSION PROCEDURE

Step 1. Define The Objective

The primary objective of a multiple regression analysis is the determination of task performance times. In a typical application, the tasks which are being measured may be part of a complete pattern of activity performed by a group or the activity may take place over a large area requiring considerable travel by work crews. At any rate the situation is such that direct assessment by a single observer, or even several observers, would be prohibitive. The analysis is performed on data which are the frequency of occurrence of work outputs, that is, completed units of activity or units of production. These units are analyzed mathematically to develop an equation which relates the unit of production (known) and its unit production time (unknown) to the total time spent at work (also known). It is emphasized that the objective in this case is the estimation of the unit times since multiple regression will be used again in Chapter 5, in the development of standard data. In that case, the objective is the prediction equation and the total time is the unknown quantity.

Step 2. Establish Activity Elements

There are a number of common characteristics between Multiple Regression and Work Sampling Systems which should be discussed. Both techniques describe the work as it occurs; neither is prescriptive of what *ought* to occur. One major difference lies in the characteristics of the activity elements. Recall that in Work Sampling, activity elements are units of work which constitute inputs to production (percent of time which machine is operating, percent time operator is waiting, etc.). In Multiple Regression the unit is some production output (number of holes dug, number of pumps overhauled, etc.). Therefore, the activity elements are not observed work behaviors; they are the countable results of that behavior. In fact, the unit times are inferred on the basis of an assumed relationship described mathematically as a multiple variable linear model. To make this distinction clear we often refer to the MRS activity as "pseudo-measurement" to emphasize that there is no direct observation of work.

If we view multiple regression in general as a means for identifying sources of variation, then the basic approach of an MRS is to describe the way in which the time spent in production (dependent variable) varies as a function of the level of output of the various activity elements (independent variables). The link between the various activity elements and the total time is, of course, the elemental time per unit. The means for synthesizing these elemental times is provided by the MRS, and the elemental times are the coefficients of multiple regression for each activity element.

Consider an example. Suppose we wish to analyze the work of custodians and janitors as part of a program of indirect cost control. We have chosen to begin with routine room cleaning operations. Records show that a custodian cleaned units **A**, **B**, and **C** in 4, 5, and 7 hours, respectively. A simple estimate of cleaning time would be the average time of 5.33 hours per unit. Scheduling custodians on the basis of 5.33 hours per unit would overestimate the time to clean unit **A** by 1.33 hours and **B** by .33 hours, and underestimate the time for **C** by 1.67 hours.

To pursue the example further, let us assume that this level of error is unsatisfactory. This would initiate a search for sources of variation. The question to be answered is, why are the times to clean the units different? One element contributing to variation in cleaning time might be the number of rooms in each unit. Upon checking, we find that **A** has twelve rooms, **B** has eighteen, and **C** has thirty. If we again apply the averaging principle, we might produce our revised estimates using 16 hours per sixty rooms or .27 hours per room. Using number of rooms cleaned as our activity element then, the estimates of cleaning time would be 3.24 hours for **A**, 4.86 hours for **B**, and 8.1 hours for **C**. The respective errors in estimated cleaning time would then be .76, .14, and 1.1 hours. This might be a more acceptable level of error; however, two assumptions must be made in order to accept .27 hours per room as the basis for our scheduling efforts: (1) all the custodial time is spent cleaning rooms, and (2) cleaning time is independent of any other characteristic of the units. We will carry our analysis one step further.

The next level of analysis begins by simply plotting the time to clean a unit as a function of the number of rooms per unit. Figure 3-1 is such a plot. One look at the plot should bring at least one of the stated assumptions into question. Note in Figure 3-1 that the line which describes the relationship between rooms and custodial time does not pass through the origin. This suggests that approximately two hours of custodial time cannot be attributed to the independent variable which was selected. The curve

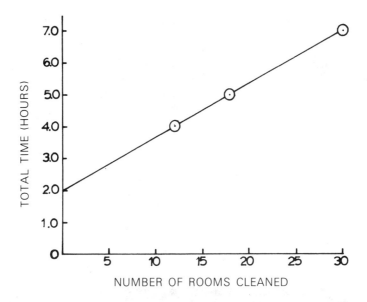

FIGURE 3-1 Relationship Between Rooms Cleaned and Cleaning Time

suggests that custodians would spend two hours even if no rooms were cleaned. That two hours might be attributed to make ready time, personal time, and/or another activity element not yet identified. If we solve the foregoing example using simple linear regression to fit a line rather than visually estimating the line, we can derive a unique solution. For this case, the equation would be:

$$\text{Cleaning time} = 1.93 \text{ hours} + .167 \frac{\text{hrs}}{\text{room}} \times \text{no. of rooms.}$$

Estimation of cleaning times for each unit and computation of the error is left to the reader.

The error in estimation is now quite small, but having two hours of a custodian's time for a unit as an unexplained constant may not be satisfactory. The logical extension of the line of reasoning to this point would be to install an MRS. Actually, the basic difference between the simple regression illustrated in the example and MRS lies in the number of explanatory or independent variables which are included in the analysis. An MRS, for example, might include the number of rest rooms and the number of square feet of hall space as additional activity elements. One such analysis of custodial work yielded the following expression:

$$\text{Cleaning time} = .83 + (.20)(\# \text{ rooms}) + (.28)(\# \text{ rest rooms}) + .000083 \, (\# \text{ ft}^2 \text{ hall}).$$

Notice, MRA now provides the following unit times for three activity elements:

> Cleaning rooms = .20 hours per room
> Cleaning rest rooms = .28 hours per rest room
> Cleaning halls = .000083 hours per sq ft
> Constant or
> Unexplained time = .83 hours per job.

We can see from the example that when the activity elements which constitute the independent variables in a multiple regression are carefully chosen, the analysis produces coefficients in the regression equation which are actually the amount of time required to accomplish one unit of that activity; that is, one unit of production. The term, *carefully* chosen, is not selected lightly because interpretation of the results of a multiple regression analysis in terms of the meaning of the coefficients lies in the realm of the art of analysis and is not readily communicated in a discussion of technique or procedure.

The activity elements for a multiple regression analysis can be characterized as follows:

Easily Enumerated. The data which are used in developing the model are simply counts of the frequency of occurrence of each activity element. The multiple regression technique requires that each frequency can be known without error. As a result, the more routine or mechanized the enumeration process is, the simpler the measurement task and the less susceptible it is to error. A key characteristic of an activity element, then, is the degree to which the element conforms to or is a part of existing information systems. (Is it currently being counted?) Short of that, the analyst should consider the extent to which the element lends itself to the establishment of a routine or institutionalized data collection system. The elements must represent countable activities.

Mutually Exclusive. The requirement for error-free specification of the independent variable makes it critical that an instance of production be recorded as an occurrence of one and only one activity element. This means that the activity elements must be unique, countable, events.

Substantially Inclusive. There is no requirement for collectively exhaustive categorization, as was the case in work sampling. The form of the regression equation is such that any time which cannot be accounted for by virtue of the enumerated activity elements will automatically become part of the constant term in the total time equation. The constant is, of course, effectively the miscellaneous category since it contains all unexplained variance. As a result, the output of the multiple regression is necessarily, collectively exhaustive. Even though the number of activity elements which the analyst specifies is largely dependent upon the use to which the results are to be put and the ease of enumeration, failure to account for the sources of variation will produce unacceptable results. The ultimate question involves the amount of time which will remain unexplained by the study. This means that the rule used to determine the number of activity elements for an MRS is much more liberal than the rule for work sampling. The analyst is encouraged to specify every activity element which reason suggests is a component of the task, and which can economically be enumerated. The use of a procedure such as stepwise regression, forward regression, or maximum R^2 (which are available in virtually every computer library routine for multiple regression) will provide the analyst with a systematic method of selecting the number of independent variables which should remain in the final equation. The details of the selection process will be illustrated in the example provided in the Computations section. Also, the potential for error will be discussed in a subsequent section.

Step 3. Determine The Scope

The basic issues in determination of the scope of a multiple regression analysis are similar to those discussed in Chapter 2 for Work Sampling studies: representativeness, unexplained variance, and cost.

Representativeness. The reader is referred to the discussion of representativeness in Chapter 2, Work Sampling. Data must be collected over a sufficient period of time to promote confidence in the generalizability of the results of the study. In the case of multiple regression studies, the routine nature of the data collection process and the tendency to have the enumeration institutionalized makes it less susceptible to an error of non-representativeness. Once the data become part of a formalized information system, the computer analysis to verify estimated unit times can (and should) be run periodically to provide continuous re-evaluation of the time values.

Another test of representativeness involves the range of variation in the frequency counts which are taken. In the example provided, the analyst should determine whether twelve to thirty rooms is a representative sample of custodial tasks. If, for example, jobs as small as eight rooms are part of the normal range of custodial jobs, then some eight-room data should be included in the MRS.

Unexplained Time. The proportion of total time which can be accounted for by the MRS is a function of three factors: (1) the ability of the analyst to select those activity elements which actually describe the work content and which affect the time spent by the worker, (2) the variability in the time required to complete one cycle of an activity element, (3) the amount of time which workers spend in activity other than that described by the activity elements. The objective is, of course, to account for 100 percent of the time spent. This does not mean that the constant term should be zero. In fact, unless the hours are discounted for Personal Time, Fatigue, and Unavoidable Delays (see Chapter 4 for a complete discussion) we would expect that time to appear as a component of the constant term. In addition, any set-up or make-ready time which accompanies the work will appear in the constant. The judgment of the relative amount of total time which legitimately belongs in a constant term should be made by persons who have full knowledge of the task requirements and job conditions.

The second criterion which will assist in assessing the contribution of the analysis to the reduction of variance is the coefficient of multiple correlation, R. (A brief explanation of coefficients of correlation appears in

Appendix B.) In most cases, one can use R^2, as an estimate of the proportion of the variance in the dependent variable which can be explained by the multiple regression equation. For example, given a study with $R = .75$, the analyst could say that the activity elements used in that analysis accounted for approximately 56 percent of the total times to perform the work ($.75^2 = .56$).

In general, the analyst seeks to conduct a study of sufficient scope to achieve a small unexplained variance. Since the standard error of the estimate is an inverse function of $(n-m)$, where $n =$ the number of observations and $m =$ the number of terms in the regression equation, reduction in unexplained variance can be accomplished by increasing sample size. Mathematically, an analysis can be conducted with $m+1$ observations; however, unless the condition is extremely invariant, the probability of achieving any kind of statistically significant results with such a small sample is quite remote.

Cost of Study. Compared to the other measurement techniques which are described in this text, MRS has the smallest direct labor component of cost. Once the tasks which serve as activity elements have been identified, the incremental cost of collecting the frequency of occurrence is usually nominal. Actually, since much of the information is already collected for inventory control or cost accounting purposes, the incremental cost may be zero. Where special data collection procedures are instituted, however, that cost must be included in the cost of the study. The cost of computer time to actually solve the regression equation is the other cost component. Most library routines are extremely efficient and the cost is low. For example, each execution of the program to solve the example problem which is discussed in the Computation Section, involving seven activity elements and fifty six observations was about .20 minutes per run. The number of runs required to achieve the level of precision required for the analysis will depend on the level of control already present. It will probably be at least six or seven runs.

Step 4. Establish The Data Collection System

The matter of organizing an information system to collect the data necessary to perform the regression analysis has already been mentioned several times. Once again, the ideal application makes use of existing information systems. One of the critical decisions which confronts the analyst is whether or not to use self-report systems to collect the data. There is a strong inclination, especially when analyzing crews or individuals who cover a lot of territory, to have them maintain the frequency counts. Since the accuracy of the information needed for an MRS are readily verified, there are fewer problems with self-report in this type of application than there are, for example, in programs to record quality problems. However, the analyst should not make the mistake of assuming that verifiability is a substitute for verification. If self-reporting is used to generate the data for the multiple regression analysis (or for any work-measurement application, for that matter) a highly visible and well-publicized procedure to sample and verify the accuracy of the data is a prime requisite. In the MRS, as was the case with work sampling, the immediate supervisor is in the optimum position to control the quality of the data-collection effort, even though with MRS he or she may not be directly collecting the data.

Part of the design of a data collection system may involve the creation of data collection forms. Whereas data forms in work sampling studies were used by trained observers, MRS forms are typically completed by operators as part of their standard operating procedures, or as an additional chore specifically added for the study. In either case, attention to the human engineering of the form to promote ease of response and error-free reporting will do much to enhance the quality of the study. Design of special purpose data collection forms should be a conscious act, not an accident. In addition, a training program to initiate the workers with the purpose of the study and the correct manner for completion of the forms is a must.

Step 5. Notify The Appropriate Parties

Since data collection can take place via existing information channels, multiple regression is possibly the most unobtrusive of all work measurement techniques. Consequently, there is a strong temptation to use it as a surreptitious means of measuring performance. Putting the general question of ethics (which is discussed in Chapter 1) aside, such an application will probably produce disappointing results. In order to extract times which are actually sensitive to the differences between activity elements and reflective of true task times, the unexplained variance must be minimized. The single best way to accomplish this is by informing operators that they are being studied and then re-enforcing that with close supervision during the period of the study. In fact, many practitioners choose to institute a program of Short Interval Scheduling in conjunction with a multiple regression analysis to provide maximum control.

Short interval scheduling is a method of work control which can be effectively used with a Multiple Regression System in the development of standard times. The technique involves decentralized forecasting and control of work assignments. In most applications, first-line supervisors would be provided with the unit time estimates which had been developed through the MRS. In the example developed earlier in the chapter, the supervisor might be encouraged to make work assignments for one floor at a time, or possibly a portion of a floor. The key is that the time estimates be used to estimate completion time for a short period of activity. The assignment is made and the expected completion time specified. Employees are instructed to operate on an exception basis. If the allotted time is insufficient, the supervisor will expect some explanation of the problems which were encountered. The objective is *not* to get people to work *faster*. The objective is to systematically review progress and identify causes of inefficiency or sources of interference. Experience has shown that continued application of the MRS during the Short Interval Scheduling period will usually reduce the magnitude of the constant term in the regression equation, increase the coefficient of multiple correlation, and improve the accuracy of the regression coefficients as estimates of unit task times. For a more complete discussion, see Richardson (1).

Step 6. Conduct The Study

Conduct of a multiple regression analysis requires less in terms of time and resources than any other work measurement technique. The single most important activity by the analyst is the periodic monitoring of the counting activity to verify its accuracy.

Once sufficient data are available to perform the computations (that is, $> m+1$ observations) it is often instructive to run a trial regression analysis. This will permit some preliminary analysis and possibly disclose the existence of some systematic problem in data collection or the failure to identify an important activity element.

SUMMARY OF THE PROCEDURE

A Multiple Regression System, which provides estimates of the time per unit for activity elements which usually occur in combination, requires four items of information: (1) a clear statement of the purpose of the MRS, (2) specification of the activity elements for which unit times are being sought, (3) an assessment of the scope of the study, (4) counts of the frequency of occurrence of the activity elements.

SOURCES OF ERROR

The errors in a Multiple Regression System can be subdivided into classes.

Modeling Error

Modeling Error is present in all multiple regression studies. It is present because the disaggregation or "sorting out" of total time spent into the time dimensions of the various activity elements is a function of the analytical model chosen for the analysis. In the predominant number of applications, the analyst uses a least-squares methodology and the data are fit to a first-order or linear equation. As a result, two factors which are likely to be present in the actual work situation may produce error. They are (1) non-linearity of relationships between the states of certain activity elements and the time to perform them, and (2) lack of independence in terms of the effect of certain combinations of activity elements.

Nonlinearity occurs when the time per unit changes over the range of values considered for the independent variable. For example, the time to fill some containers might not be a linear function of the height to which it is filled. The time would most likely be a linear function of the volume to be filled.

Lack of independence is a more difficult problem. Sometimes, the time to accomplish one unit of an activity element is dependent on how many units of another element must also be completed. In this case the times are said to co-vary. Unless the analyst is very familiar with the activity being analyzed, errors can occur. Sophisticated analytical models are available to handle these conditions. They are beyond the scope of this discussion.

Another, more subtle problem in independence is often disregarded by MRS practitioners. The analyst must be aware that the study has been conducted on a specific work milieu. A pattern of activity elements exists. For example, if after the study one worker is permitted to specialize in performing a single element, the time per unit computed from a nonspecialized pattern of work would most likely be incorrect. The error, then, is due to the fact that an MRS analyzes the work system as it exists. The extraction of a single unit time for one activity element from all the other unit times violates the wholeness principle in the system. When implementing an MRS, the analyst frequently violates this constraint. That action should be taken knowingly and the analyst should exercise discretion when using single element unit times. This is especially true if the elemental times are used in conjunction with Short Interval Scheduling. There may be a tendency to treat the times as being more precise than they actually are. Failure to accomplish a task in a unit time can be due to the inappropriateness of the unit time value, but the fact that the times are "computer generated" often makes it difficult for some supervisors and managers to accept their somewhat qualitative nature.

Representativeness

The errors of Representativeness in MRSs are, in principle, closely related to those discussed for WSSs. Sometimes nonrepresentative selection of values of the activity elements will occur. In studies performed on data of known correlational properties, it has been shown that omission of extreme values (very high and very low) will cause the computed coefficient of correlation to be lower than the actual one. If intermediate values are systematically rejected, an incorrectly high coefficient of correlation will be computed.

Occasionally the dependent variable may also be edited or selectively reported. For example, one might be tempted to eliminate data for days in which an unusually high or low number of hours were worked. If this is actually nonrepresentative of the actual work situation, the accuracy of both the coefficient of multiple correlation and the coefficients of multiple regression will be affected. The latter, of course, means inaccurate elemental time values.

A third source of nonrepresentative analysis is encountered through the statistical analysis which usually accompanies a computer multiple regression analysis. Forward, stepwise, or maximum R^2 procedures, among others, are intended to provide the analyst with a basis for selecting a model which will efficiently predict the state of the dependent variable. Unfortunately, these models which use a subset of the actual activity elements tend to overestimate the coefficient of correlation. Since the MRS seeks coefficients for all appropriate activity elements, use of a reduced set of elements may be inappropriate. However, occasionally an analyst becomes more interested in the statistical sophistication of the computer analysis and loses sight of the work measurement goals.

In each of the conditions cited above, the error was introduced because the data used for analysis did not represent the actual work conditions. In order to protect against this type of error, it is recommended that the analyst scan the raw data. It is important to observe the range of values for each activity element to assure that legitimate but extreme conditions are represented. Beyond that, the values should generally conform to the experience of those who know the range of activity. If the data are suspect, however, they should not be altered. The actual source of the problem should be investigated.

Sampling Error

Sampling Error is present in MRSs just as it is in every measurement system. This happens because we are attempting to generalize the results of a specific set of observations to a much larger set of events which almost always occur in the future. In the case of both Multiple Regression Systems and Work Sampling Systems the precision of the time estimate is essentially the inverse function of the number of observations. Since the data collection is such a low-cost matter in most MRS applications, it is simply not appropriate to tolerate large sampling errors.

COMPUTATIONS

The computations which are required for a Multiple Regression Analysis for more than a few variables and a limited data set are extremely laborious when performed manually. In fact, unless the analyst has access to a computer, a work measurement program using MRS to any extent should simply not be attempted. The several forms of the multiple regression equations are included in Appendix C since they are required as part of Chapter 5—Standard Data. In Chapter 5, also, the computations for a small data set are illustrated. In the example which follows, primary attention is paid to computer-generated outputs and their application in a Multiple Regression System instead of the development of work standards.

The data for this example were collected in a large metropolitan department store. The workers who were involved operated in a clerical pool and both the size of the pool and the activity of the individual clerks varied by day. The activities which here made up the bulk of their productive output were as follows:

Y = Man-hours
X_1 = Number of pieces of mail processed (open, sort, etc.)
X_2 = Number of pieces of miscellaneous mail processed on an "as available" basis
X_3 = Number of change order transactions processed
X_4 = Number of window payments (customer charge accounts) transacted
X_5 = Number of checks cashed
X_6 = Number of money orders and gift certificates sold
X_7 = Number of bus tickets sold (exact change required on bus)

Table 3-1 shows fifty two days worth of data used to perform the first Multiple Regression Analysis. This particular unit had not been accustomed to any type of work measurement and was staffed on the basis of experience.

TABLE 3-1
UNITS OF PRODUCTION AND TOTAL TIME FOR A DEPARTMENT STORE CLERICAL UNIT

OBS	DAY	Y	X_1	X_2	X_3	X_4	X_5	X_6	X_7
1	T	116.8	9964	91	51	1007	570	147	670
2	W	103.3	8968	100	52	1093	515	243	936
3	R	133.3	9306	56	54	1525	1007	249	755
4	F	113.0	2727	93	52	1310	817	286	1335
5	S	91.3	6247	14	45	913	463	361	305
6	M	108.4	5121	131	59	1281	489	216	835
7	T	99.5	15080	112	55	713	328	204	305
8	W	116.8	7381	71	52	850	421	101	875
9	R	119.3	9466	157	57	995	749	137	605
10	F	92.0	10372	115	79	779	481	180	525
11	S	79.7	6065	105	35	496	382	147	195
12	M	113.3	5554	100	67	1021	490	80	440
13	T	94.3	3943	88	54	648	354	113	415
14	W	94.2	9313	47	49	939	389	74	825
15	R	121.1	5239	106	61	1204	887	129	865
16	F	101.3	4341	64	70	1087	671	178	1315
17	S	88.8	8001	53	38	635	465	138	400
18	M	135.1	9536	76	58	1193	479	109	1030
19	T	123.5	9091	73	50	646	354	78	415
20	W	119.4	10289	105	71	806	424	109	880
21	R	136.7	7144	76	72	646	646	140	600
22	F	113.5	5555	71	77	1023	519	72	505
23	S	86.2	8829	17	38	502	442	74	185
24	M	103.7	4647	121	52	732	468	97	710
25	T	80.3	6594	84	54	442	368	78	215
26	W	106.9	6711	100	50	696	440	139	560
27	R	116.5	2736	63	73	1201	933	227	1100
28	F	104.0	5621	64	74	1107	990	122	1320
29	S	99.5	6347	22	47	890	856	224	560
30	T	108.1	9090	119	57	1612	572	234	870
31	W	111.1	9395	117	47	1292	424	149	880
32	R	129.8	10629	105	68	1075	734	120	654
33	F	121.7	6261	79	63	852	476	134	445
34	S	91.1	8077	74	45	524	393	83	310
35	M	106.2	8846	132	52	982	512	103	820
36	T	88.5	6844	106	44	686	323	80	115
37	W	89.1	8956	54	49	811	436	67	660
38	R	124.1	8683	71	68	1042	768	107	1000
39	F	101.5	4024	97	64	919	665	210	1565
40	S	77.5	7348	25	48	621	418	79	457
41	M	105.6	6314	159	65	1013	444	81	670
42	T	99.0	8325	63	43	731	297	81	430
43	W	109.6	7271	86	48	941	347	67	985
44	R	119.8	5915	91	61	883	797	103	765
45	F	102.1	2346	113	65	712	534	99	570
46	S	84.1	6973	48	39	491	421	76	285
47	M	109.1	5280	77	39	491	507	96	595
48	T	102.0	7535	81	29	1206	354	63	255
49	W	102.7	7049	41	33	639	407	53	740
50	R	126.1	3691	86	40	683	762	87	825
51	F	111.0	3067	81	28	724	710	126	1515
52	S	73.4	8764	12	34	975	508	147	558

In order to demonstrate the principle of removing variance through multiple regression, a simple regression (one-variable) analysis is shown using the activity element (X_5) which explains the most of the difference in hours recorded.

$$\text{Hours/Day} = 84.2 + .0399 \text{ (\# Checks Cashed)}$$

The statistical analysis of the one-variable model shows then $R^2 = .237$, suggesting that not quite one-fourth of the variability in total man-hours worked can be explained by its relationship to the pattern of check-cashing activity. The t-test for significance of the estimate of about .04 hours per check to cash checks gives $t = 3.94$, $\alpha \leq .0005$. This would indicate that on an average day (543 checks cashed) about 21.7 man-hours is spent cashing checks. Since check cashing is only one of nine activities analyzed, a practical question arises, why does the number of checks cashed become the best predictor of work load? A review of the data reveals that the number of checks cashed varies from 297 to 1007. The pattern shows Thursday and Friday to be heavy check cashing days and also days in which more man-hours are generally recorded.

However, the more significant equation from an MRS standpoint is, of course, the seven-variable model. Table 3-2 shows each activity element, its regression coefficient and the significance level for the t-test. The coefficient of multiple correlation is .67, giving an R^2 of .45. This indicates that less than half of the time spent by employees in this unit is accounted for by the seven activities listed. The constant term of 57.4 hours, compared to an average of 106 man-hours per day, bears this out.

TABLE 3-2
RESULTS OF A MULTIPLE REGRESSION ANALYSIS OF A DEPARTMENT STORE CLERICAL UNIT

Activity Element	Hours/ Unit	Average Units/ Day	α
Number of Pieces of Mail Processed (X_1)	.00095	7153	.23
Number of Pieces of Misc. Mail Processed (X_2)	.095	83	.10
Number of Change Order Transactions (X_3)	.170	53	.28
Number of Window Payments Transacted (X_4)	.011	890	.23
Number of Checks Cashed (X_5)	.034	543	.01
Number of Money Orders Sold (X_6)	(−).056	132	.08
Number of Bus Tickets Sold (X_7)	.0056	686	.59
Constant (unexplained variance)	57.4		

The presence of a negative coefficient for X_6 is understandable from a statistical point of view. Psychologically, however, it falls in the realm of a perpetual motion machine. Several possibilities present themselves. First, since the significance level of all but three of the variables is not spectacular, more data are needed and should be collected. Second, X_6 might be combined with checks cashed (X_5). Third, and probably the most crucial is an effort to manage the activity of the unit, based on the insights gained from this analysis. For example, it takes one hundred times as long to process a unit of miscellaneous mail as it does "regular" mail. Specialization—dedicating one clerk to miscellaneous mail might work. There is sufficient average load (83 pieces × .095 hours/piece = 7.88 hours) to warrant this action. Another possibility is the dedication of certain personnel to servicing customer window calls (X_4, X_5, X_6, and X_7), especially on Thursday and Friday seems warranted. More solutions present themselves to those who are closer to the situation; our purpose here is simply to show the application of MRS. Until some effort to manage the size of the unit and increase productivity, the existence of a negative coefficient for X_6 should not be a source of concern.

SUMMARY

The technique of using multiple regression to develop standard times for work is the least obtrusive of all measurement techniques, since direct observation of the worker is not necessary. The primary applications of this method have been in the development of standards for indirect labor activities. When circumstances permit, MRSs have been combined with Short Interval Scheduling to accomplish control of maintenance, clerical, and similar activities. The full worth of MRSs has yet to be realized. It is one of the most fertile of all measurement approaches for research and creative application.

REVIEW QUESTIONS AND PROBLEMS

1. What are the characteristics of activity elements for Multiple Regression Systems?
2. What factors must be considered in determining the scope of a multiple regression program?
3. Explain the major sources of error in Multiple Regression Systems.
4. In using a multiple regression approach to establishing standards, how would you decide whether or not to use the regression equation which you develop in actual application?
5. Why are programs such as short interval scheduling normally incorporated into a multiple regression program?
6. In a purchasing unit, the accounting clerks process two types of purchase requisitions, the short form and the standard form. Estimate the processing times for both using multiple linear regression.

Total Minutes	No. of Standard Forms	No. of Short Forms
302	26	34
290	25	35
315	26	35
305	25	36
333	26	37
392	26	38
336	25	37
337	26	39
339	25	41
340	25	43

REFERENCE

(1) Richardson, W. J., *Cost Improvement, Work Sampling, and Short Interval Scheduling*, Reston Publishing, Reston, Virginia, 1976.

4

TIME STUDY

A Time Study System (TSS) uses direct observation time study to determine the time which should be allowed a worker, who has normal skill and ability and is working at a normal pace, to perform a defined task according to an approved method, and under specified conditions.

THE TIME STUDY PROCEDURE

Step 1. State The Objective

In applying direct observation time study the objective is essentially implicit in the selection of the technique. In the words of the definition, the analyst's objective is to determine "the time which should be allowed . . . ." This is the first example in our text of measurement which produces normative rather than descriptive results. It is also the single most used technique for measuring work in industry today. By and large, application has been predominantly in the private sector, but recent concern over low productivity, high cost, and a felt need for standards of work output has been generating increased use of TSSs in the public sector.

The analyst should be explicit in the selection of a TSS by acknowledging that he or she is seeking the normative standard. The relationship between the concept of normal used in performance rating and the anticipated level of incentive earnings should also be recognized.

This is necessary because there is a close relationship between the concept of normal used in performance rating and the expected performance of the workers, as compared to the normative standard. Both normal pace and expected performance are covered in detail later in this chapter. The matter of the normative nature of the process is raised here because we seek full realization by the analyst that TSSs are prescriptive and thus are subject to complications which are not present in other measurement techniques.

Step 2. Notify The Appropriate Persons

Notification was discussed in Chapter 2, but with respect to time study there are two additional reasons for notification which make it important as an early step. First, the analyst must determine whether the task is in a stable condition; that is, whether the design, the materials, conditions, etc. can be considered representative of the task as it will exist for a reasonable period of time in the future. Second, the analyst must verify that there are operators performing the task who know the proper method and who have sufficient skill, ability, and practice for the job.

In order to accomplish the multiple objectives and obtain the information needed, the persons who are notified should include the first-line supervisor, the union shop steward and/or time study representative, and the operator, once one is selected.

Step 3. Determine The Scope

As is the case in every measurement technique, the scope of the study must be viewed in terms of representativeness and errors.

Representativeness. Time Study is unique among the measurement techniques with respect to the representativeness issue. The problem, of course, is whether or not the conditions present when the study was taken are typical of the conditions which will be present when the time standard is in effect. Every act, selection of the operator, analysis of the method, and recording of job conditions is actually designed to insure representativeness. However, once the study is completed and a time standard is developed, the documented time study takes on a new importance. There is usually a trial period of thirty to sixty days during which either labor or management can request additional study of the job. After that period the time standard becomes permanent. If a condition arises in which the time standard appears to be inappropriate, the analyst reviews the conditions surrounding the job. If the operator is qualified, all other conditions are compared with the documented time study sheet. If the existing conditions differ with those recorded on the time study, those on the study sheet are considered fact. The discrepancies are dealt with as changes which have occurred since the study was conducted.

Errors. The problem of errors and their effect on time study results is discussed in detail later in this chapter. However, in the context of the scope of the study, the decision regarding the number of observations which should be included in the sample is an important one.

There are, of course, standard statistical computations which can be used to provide some desired level of confidence that the true average time for the activity element is satisfactorily close to the average time computed from the study. The statistical analysis is developed in Appendix A. In the section on computations, an example statistical analysis is illustrated. However, in actual practice, statistical criteria are rarely, if ever, applied. The fact of the matter is that most analysts have little intuitive feel for the variance of the elemental times, and thus are not in possession of the data necessary to make *a priori* computations to estimate the number of cycles to be studied. The philosophy of the practicing analyst seems to be that the fixed cost of preparing for the study is so high that an appropriate strategy is to study as many cycles as possible. Rules of thumb suggest that no fewer than twenty cycles or thirty minutes of observation should be used for a single study. Due to the way work is organized, some studies can require full shift or more to record the relevant time data.

A word of caution is offered to analysts. As union representatives become more and more sophisticated, the quality of their grievances with respect to incentive rates is improving. In the future, it may not be unusual for a grievance to be argued on the basis that an insufficient number of observations was obtained in the study to provide a requisite level of precision, or that the sample was not representative of the actual work activity. Statistical analysis of time study data may soon become routine procedure rather than the exception.

The question of cost as a function of the length of a time study has already been mentioned. In the case of TSSs, the analyst designs, performs, and analyzes the results of the study. As a result, the direct influence of cost considerations on the scope of the study requires little more comment.

Step 4. Select The Operator

It is, of course, imperative that the operator use the approved method while being studied. In addition to operator knowledge and ability, skill is an important factor. Care must be exercised so that the operator who is studied is *not* exceptionally skilled. Performance rating will be explained as part of step 6, but it should be mentioned here that when studies are made using operators with high skill, it is often difficult for the analyst to rate the operator's performance correctly. Even if the performance rating were correct, the other operators are usually apprehensive concerning their ability to perform against the standard which has been developed from a study of a "superstar."

With respect to the pace which the operator chooses during the study, experience shows that the fairest standards are developed when the performance of the operator during the study is close to that which he or she normally exhibits during work. In this regard, it is a duty of the time study analyst to create an environment and a relationship with the workers which is a cooperative one. Both labor and management benefit when a standard is fair; both suffer when it is not.

Step 5. Establish Activity Elements

It is possible to perform a time study by simply recording the total cycle time per unit of output. However, a number of considerations make it not only desirable, but imperative that the time study be performed on an elementally subdivided task, not the entire cycle. First, knowledge of time for specific activity elements can indicate where efforts to improve the method and reduce time per unit might be successfully applied. Second, those elements which might be externally paced (machine controlled or conveyor paced elements) must be separated from those which are under operator control. This is especially important for multiple machine loading problems and incentive pay applications. Third, occasionally an operator will exhibit substantially different levels of performance on some elements of the task. Elemental breakdowns permit the time study analyst to compensate for this. Fourth, most collective bargaining agreements with unions provide that, when methods are revised, only those activity elements which are directly affected by the change may be revised for the purpose of computing new standard times. Fifth, without elemental time values, it is impossible to develop standard data (See Chapter 5).

The activity elements which are appropriate to a time study possess the following attributes:

Logic. Elements should make sense. That is, an element should consist of a group of related motions. Whenever possible, activity elements from different jobs should consist of the same motions, as this is one of the keys to good standard data development.

Discreteness. Elements should have easily recognizable beginning and end points. Typically, good elements end with marked body or limb movements. Better still, they might generate a characteristic sound, making it possible to detect the end point without a visual fixation.

Duration. Elements should be short enough to provide discrimination of the activity which makes up the job. Conversely, they must be long enough to be accurately timed. A rule of thumb is that no element should be less than .05 minute duration. Occasionally shorter elements can be timed, especially if they are followed by a long element. This permits the analyst to observe two end points and then record both times during the long element.

Control. Machine controlled elements should be separated from operator controlled elements. This is particularly critical if standards are being developed for incentive applications. The problem arises because an operator can only affect the time of certain elements. The effort expended on those elements does not influence that portion of the cycle which is machine controlled. Therefore, machine controlled elements are given special treatment in the performance rating.

Step 6. Conduct The Study

There are three basic functions which make up the actual conduct of the study. They are the recording of job conditions, the recording of time to perform the task, and the rating of operator performance.

Record Job Conditions. Figure 4-1 is an example of the back of a time study form. In addition to precise descriptions of the activity elements, notice the wealth of data requested of the analyst. This includes a sketch or photograph of the work area with appropriate dimensions. The information gathering rule is simple; identify any factor which might affect the level of performance observed during the time study. Very often, a change in conditions surrounding a task will cause productivity to be substantially above or below that determined by the study. If the conditions have been recorded, the assignable cause is easily identified and appropriate correction made. If the assignable cause has not been captured, the analyst will suffer a loss of credibility (and possibly employment).

TIME STUDY

Timed By	G. L. Smith	Checked By	W. J. Richardson	Workplace or Mach.	Mach. No.	12F
Operators Name	C. Sink	Clock No.	12371	Material 1060 Steel	SFPM R.P.M.	5 m
Time Study No.	75-34	Dept. No.	4	Lubricant Heavy duty soluble oil (10)	Strokes Per Min.	
Special Tools Used	Sintered Carbide				Feed	.02 ipr
Part Name	3" OD Shaft	Part No. 31255 27		Operation No. 37B		
Remarks						

SKETCH — **DETAILED DESCRIPTIONS OF ELEMENTS**

A. Load Chuck. Select Shaft, Return 30" to lathe, Place in chuck and tighten, aside key
B. Advance tool to appropriate depth, engage feed.
C. Turn O.D. per attached spec — machine controlled.
D. Retract tool, return to rest position
E. Unload Chuck, remove chips, remove shaft, aside shaft 24"
F. Walk 10 ft. Get batch of 5 shafts, return to machine.
G. Reach 12" get Micrometer, move 12" to shaft, measure OD, adjust depth of cut

Note: tool change by set-up operator

FIGURE 4-1 A Time Study Form

Record Time To Perform Task. This is the step in which the analyst actually times the operation. The objective is to accurately record the amount of time which the selected operator uses to perform each activity element during the period of the study. In most cases, this is accomplished by the continuous-reading stopwatch method, with the elapsed time between successive readings being the observed time for that activity element. The alternative method, snapping back the watch for each element and simply recording elemental times, is generally unacceptable to labor since the opportunity for cheating by the analyst is greater. Since the general approach to work measurement should maximize the sense of objectivity and fairness, the choice of method should not be made a matter of contention.

Upon arriving at the work place, the analyst should engage the operator briefly in conversation. The analyst must be certain that working conditions are representative. If the operator mentions any discrepancies, these should be recorded on the data sheet for future reference. Again, the analyst should encourage the operator to perform at the pace to which he or she is accustomed. More will be said about this aspect of the study, but the value to be gained is extremely difficult to communicate to the operator.

The analyst should take a position out of the visual field of the operator but in full view of the operation. This is usually to the rear and one side of the operator. Care must be taken to not be so close as to interfere in any way with the performance of the task. Modern practice suggests that the analyst should consider sitting to time study a seated worker.

At this point, the analyst is ready to begin recording the observations. If the operator is allowed to execute a few cycles before the observer begins recording data, it permits the operator to establish his or her accustomed pace. At this point the observer should record the time of day and begin the data collection. As each elemental end point occurs, the analyst will record the time at the appropriate place on the data sheet.

It is not unusual for unplanned incidents to occur during a study (adjust clothing, dropped tool, etc.). If these are noticed, the observer is advised to circle the reading which is recorded for that activity element. These elements warrant close scrutiny during the computational phase of the study. Actually, there are three types of elements which are observed during time studies: (1) cyclical elements, (2) noncyclical elements, and (3) foreign elements. A detailed discussion of each type of element and its relationship to the total study is contained in the section on Sources of Error.

Occasionally the analyst will lose concentration and miss the end point of an activity element. When this happens, a dash should be recorded to indicate the error. At the time of computation, this will cause the loss of two readings since the end of one element and the start of the next has been omitted. If several end points are missed, the analyst should consider observing a few extra cycles to insure that there are sufficient readings.

Rate Operator's Performance. Performance rating requires that the analyst observe the pace at which the operator is working, mentally compare it to a preconceived concept of "normal pace," and evaluate the pace of the operator relative to normal. In virtually every performance rating scheme in use today, normal pace is assigned a rating of 100 percent. Operators who work slower than normal will be rated below 100 percent, operators working faster than normal, above 100 percent. If we recall that the purpose for conducting a time study is "to determine the time which *should* be allowed a worker . . . who is performing at a normal pace", the role of performance rating becomes somewhat clearer. The analyst must adjust the actual time which was observed to correspond to the normal time. Computationally this is achieved by multiplying the actual time by the performance rating factor. This will increase fast (shorter) times to the appropriate normal time and reduce slow (longer) times to normal time. With perfect performance rating, several different workers, producing at different paces, but all using the same method, should generate equal normal time estimates. In actual practice we try to hold the variability to about ±5 percent.

Some operators, unaccustomed to close, continuous observation become excited and work very rapidly during a study. Others, influenced by suspicion and the conviction that management's intent is exploitation, exhibit slow and labored behavior in an effort to mislead the analyst. Occasionally, either event will generate a performance of such poor quality that the study must be terminated since the resultant standards would bear little or no resemblance to actual conditions.

These social factors militate against the possibility of generating a good standard. Bad standards reinforce the social impediments and the process can result in progressive degeneration of the relationship between labor and management. The performance rating function is often at the center of this problem. The concept of normal and other fundamental issues are explored in the section on Sources of Error.

Activity elements may be performance rated separately or a single performance rating can be estimated for the entire study. The practice differs. Often certain elements, especially noncyclical elements, are performed at a pace substantially different from that of the cyclical elements. Machine controlled elements must be treated separately, since the operator is not in control; and no additional effort on his part can, by definition, affect the required time. The treatment depends upon the ultimate use of the standard. For estimating and day work applications, the machine controlled elements should be performance rated at 100 percent of normal. However, if the standards are to be applied to incentive pay systems, the use of machine time as normal (a performance rating of 100 percent) would mean that even though the employee *cannot* increase the output during that segment of the work cycle, his or her ability to earn above base rate would be zero. A reasonable solution to this difficulty can be reached by simply performance rating machine controlled elements at the expected earnings level for the incentive plan. This means that when the operator runs the machine at the prescribed rate, he or she will, in fact, earn the incentive increment. One side benefit which accrues here is that there is little reason for the employee to surreptitiously alter machine settings to operate at excessive speed in order to increase earnings.

Step 7. Specify The Allowances

The final task of the analyst before beginning computation of the standard time for a task is to specify the allowances which are appropriate. The allowances are actually adjustments to the normal time per cycle to account for nonproductive portions of the working day. Then, when the standard time is converted to cycles per day or number of units per day, the operator's target will be attainable during the productive period. Traditionally, those nonproductive elements for which allowance is made include *p*ersonal time, *d*elays which are not under the control of the operator, and loss of production due to *f*atigue (*PDF*).

The philosophy which underlies the use of *PDF* allowances is simply that the operator should be able to make his target quantity by working at a normal pace. If allowance were not made for *PDF*, the operator would have to work at a higher rate simply to *earn* enough time to overcome the nonproductive periods. The adjustment is made to the time per cycle, but it is not expected that the operator would experience the nonproductivity during each cycle. On the contrary, while working at a normal pace, the operator actually accumulates the extra time which is then expended at some other time during the work day.

The common practice is to express allowances as a percentage of normal time. Also, the magnitude of the allowance is a matter of bilateral agreement between labor and management. Negotiated allowances are used because only one component—unavoidable delay—is actually considered to be measurable with standard technology. (Those analysts who do measure unavoidable delays use the work sampling technique which was covered in Chapter 2.) Table 4-1 is an example of one "standardized" approach to specification of allowances.

TABLE 4-1
PERSONAL AND REST ALLOWANCES

	DESCRIPTION OF WORK	Lbs. per Load	Very Light 0-3#	Light 4-10#	Medium 11-25#	Light Heavy 26-45	Extra Heavy 46-80	Heavy 81-115	Extreme 116 & up
SIT	Using hands and full arms	%	10	11	13				
	Standing only (no load)	%	11						
STAND	Using hands and full arm	%	11	12	14				
	Using hands, full arm, feet & back	%	12	13	15	18	24	31	38
WALK	Walking only (no load)	%	12						
	Using hands, full arm and back	%	12	13	16	20	26	34	42

TRUCKING

PUSHING OR PULLING TRUCKS	Empty	to 399#	400 to 799#	800 to 999#	1000 to 1199#	1200 to 1499#	over 1500#
Two wheel barrel truck	12	17	18				
Sausage tub truck	14	17	19	22	24		
Jeep - 4 wheel	13	16	18	21	23	26	
Cellar box truck	13	16	18	21	23	26	
Lift truck	14	16	18	20	23	26	
Semi-live skid	13	16	19	22	25	28	
Sausage loaf racks	15	17	21	24	26	29	
Sausage trees	12	13	15				
Electric truck (walk)	12	12	12	12	12	12	12
Electric truck (stand)	11	11	11	11	11	11	11
Electric truck (sit)	10	10	10	10	10	10	10

ADDITIONAL ALLOWANCES FOR TEMPERATURE FACTOR

Temp.	SITTING Using hands and full arm	STANDING Standing only (no load)	Using hands and full arm	Using hands full arm feet & back	WALKING Walking only (no load)	Using hands full arm and back	IMPORTANT
100	2	2	3	4	3	6	Rate each element separately.
95			2	2	2	4	3% constant
90						2	steeling knife of cutting time.
46-89							Eliminate steeling time from Study.
45	2	2	1				
40	3	3	2				
35	4	4	3	1			
30	5	5	4	2	1		If Delay Allowance is necessary, show as separate element on Standard Sheet.
25	6	6	5	3	2	1	
20		7	6	4	3	2	
15		10	8	6	5	3	
10		13	10	8	7	4	
5		16	12	10	9	6	
0		19	14	12	11	8	
-5		22	17	15	13	10	
-10		25	20	18	15	12	

The method of determining and applying allowances in the computation of standard time is another source of measurement error and contention with labor. The issues which are pertinent to this problem are discussed in the section on Sources of Error. At this point, it will suffice to say that specification of the amount of *PDF* allowance for the task is the seventh step in the process of time study.

SUMMARY OF THE PROCEDURE

Direct observation stopwatch time study to estimate the allowed or standard time for a given task is based on five items of information: (1) a record of the state of environmental factors which might affect the results of the study, (2) the specific activity elements which make up the approved method, (3) the observed times for the activity elements, (4) the performance rating or leveling factor, and (5) the appropriate *PDF* allowances.

Day-to-day administration of the collective bargaining agreement is a matter of mutual trust. One major contributor to mutual trust is the handling of standard times and any grievances which might result from disagreements regarding standard times. These grievances can only be handled intelligently if both parties possess the factual information necessary to evaluate a standard. Relations with labor can be enhanced substantially if, as a matter of practice, the time study analyst will, *before leaving the site of the study*, provide the union representative with: (1) the total elapsed time of the study, (2) the number of cycles completed during the study (number of units produced), (3) the performance rating, and (4) the *PDF* allowance.

Whether or not the operator receives the same information can be a problem, since many operators do not understand performance rating. Your author would suggest that if the information is provided to the employee, it be done by the representative of the union.

EXPECTED PERFORMANCE

Earlier in this chapter, the concept of expected earnings on an incentive plan was mentioned that was with respect to the performance rating of machine controlled elements. There is a larger issue which must be understood by a time study analyst, and that relates to expected performance.

The psychological principle of the *hypothesis of par*, states that individuals establish for themselves some level of performance which they rarely exceed. In most cases, this level is below the level of performance the individual is actually capable of attaining. There are a variety of ways which are used to raise the level, a process usually called motivation. If we ignore the use of coercion and threats (which are almost universally rejected as motivators), industry relies on two motivators, financial rewards and knowledge of results (*KOR*).

The base condition in industry constitutes a splendid example of the hypothesis of par. It is common experience that employees who are paid *day work* (straight hourly pay) generally perform at about 60 to 90 percent of normal, depending on specific local conditions (level of supervision, type of process, environmental conditions, etc.). This means that the output will not earn the base pay—usually related to 100 percent performance.

If a work measurement system is installed, standards of production established, and feedback of actual performance compared to standard *KOR* is provided, the system is referred to as *measured day work*. Under these conditions, it is not unusual to experience performance in the 95 percent to 100 percent range.

Incentive plans add piecework pay for units in excess of the standard to *KOR* as a compound motivator. Experience here shows employees producing in the 115 to 135 percent range. This, of course, assumes that there are no other conditions which interfere with the incentive nature of the program.

Figure 4-2 was partially developed from actual experience and it vividly illustrates the foregoing discussion. Under the hourly pay or daywork (**a**) condition the average worker in this shop was producing at approximately 70 percent of normal. When the same department was put on a carefully administered incentive plan, the average performance rose to 130 percent (**c**), almost doubling the number of units produced per unit time. The standard deviation of the two distributions was approximately 10 percent yielding an approximate range of ±30 percent from the mean. The center distribution (**b**) was not developed from the actual case data, but it is provided to show the complete picture of the range of performance as a function of motivational conditions.

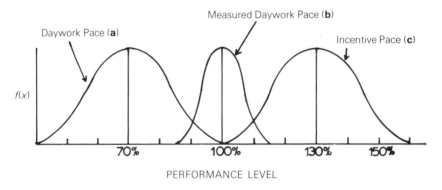

FIGURE 4-2 Three Performance Distributions

The distribution is illustrated with a standard deviation of about 5 percent. The variance is less since most measured daywork situations have a certain amount of machine or external pacing associated with the task. Note that the standard—100 percent of normal pace—does not change. The average observed performance does. This concept is often misunderstood.

One other fact of life should be elaborated. The distributions shown in Figure 4-2 are all normal distributions. In actual practice, the performances are not quite normally distributed. The tails of the distributions tend to be truncated. For example, it might be the case that no performance over 140 or 145 percent would be observed, even though the mean is 135 percent and the standard deviation about 10 percent. This occurs because there is often peer group pressure to restrict production for fear that management will restudy the job. On the low side, performance below 50 percent would not be tolerated in many daywork situations, so that distribution would most likely be truncated at about 50 percent. The other place a non-normal distribution would be observed would be on incentive programs with an average of 120 percent. We rarely observe anyone in an incentive application who consistently produces below 100 percent. This means that the tail of the distribution would be clipped at 100 percent, almost regardless of the mean.

It can be seen that the level of production which one discovers in a given situation is a resultant of a complex set of conditions. It is important to reiterate that the *standard* which is set should be measured and arrived at in an objective manner. There is a local concept of *normal pace*, but given that the remainder of the determination of standard or allowed time is ideally independent of the expectations referred to above.

SOURCES OF ERROR

Error which is introduced in a time study can originate from three general sources: (1) the equipment and its use, (2) adjustments to observed time, and (3) the operator's behavior during the study.

Equipment

The basic equipment needed for time study is a stopwatch, a clipboard on which the watch can be mounted, and a prepared data sheet. A simple sweep-hand watch calibrated in decimal minutes meets all the requirements for the task. The single most substantial improvement is the ability to stop the hand while the analyst reads to end point time value. The stop action feature permits the analyst to concentrate on the end point and will reduce the number of missed readings. There are currently three solutions to that objective; split-hand, three-watch board, and digital watch. The three-watch board and the digital watch-board are illustrated in Figure 4-3. The cost may or may not be a consideration. The analyst should also consider the total weight of the instrumentation and board in making a choice. From the standpoint of relative contribution to error, however, the error which is introduced from erroneous reading of the watch is almost insignificant.

Adjustments — Performance Rating

As was indicated in the discussion of the time study procedure, performance rating adjusts the time per cycle at the observed pace to correspond to the time per cycle which would be required at a normal pace. The judgment required is classified by psychologists as an absolute judgment. The human capacity for absolute judgments is quite limited. In fact, it appears that seven to ten categories would be the maximum number that could be expected. This would suggest that a well-calibrated analyst might be able to reliably perceive 10 percent intervals from about 70 to 140 percent. With substantial practice finer discrimination might be attained, but it would be unlikely.

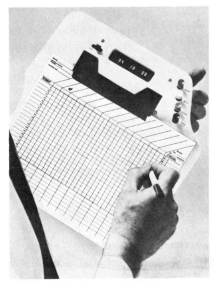 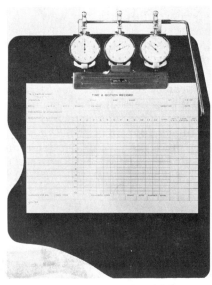

A. DIGITAL WATCH/BOARD   B. THREE-WATCH BOARD

FIGURE 4-3 Two Special Time Study Boards

Source: Reproduced with permission of Meylan Stopwatch Corp., New York, New York.

Several absolute guides to normal pace do appear in the literature. They include a person walking at three miles per hour or the pace required to deal fifty-two cards in four piles at the corners of a twelve inch square in .50 minutes. These absolute standards must be contrasted with empirical evidence that the concept of normal pace differs from company to company, industry to industry, and across geographical sections of the country. It is your author's opinion that normal pace should be viewed as the product of a cultural milieu. It actually represents a norm for a specific set of conditions.

One key to the performance rating process is obviously the training of the analyst in a concept of normal pace. Training is normally conducted through the use of standardized films which have been performance rated by a large number of practicing time study analysts. The rates which the analysts judge are corrected for differing concepts of normal and then averaged. Each set of films has a range of different answers to be used during training. Here again, the values which are used are chosen to conform to the particular organization's concept of normal.

One of the most popular systems has the trainee plot his or her estimates against the actual values for a series of different jobs performed at different rates. The results are then analyzed with respect to errors of accuracy and precision. Figure 4-4 shows some example forms with a number of different errors. Errors of precision are measured by the percent error and

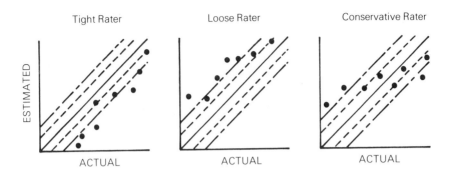

FIGURE 4-4 Three Types of Rating Error

the parallel times mark ±5 percent and ±10 percent. (Note that the axes are logarithmic, so that the percent error lines are parallel to the center line.) Errors of accuracy are grouped in three classes. Raters who constantly rate below the actual value are referred to as *tight* raters. Those above actual are *loose* raters. A more common error is called the error of *conservatism* or the tendency to rate close to 100 percent. This results in loose rating of performance below 100 percent and tight rating of performance above 100 percent. It is for this reason that analysts are cautioned about studying jobs when the operator is working at a very slow or very fast pace.

Machine controlled elements can also be a source of error. Since the operator is actually paced by the machine, he or she will not be able to alter that portion of the cycle by additional effort. Remember, the practice is to rate the machine controlled at the *expected* level of performance under an incentive application. For other applications (day work or measured day work) a 100 percent rating is used.

Adjustments—Allowances

The adjustment to normal time made for *PDF* allowances can introduce error in two ways, the magnitude of the allowance and the way it is applied computationally.

We have already mentioned that the delay may be measured using work sampling, but fatigue and personal time are negotiated. In the special conditions of space flight where astronauts are wired for physiological measurement, the personal and fatigue factors can be determined precisely and, in fact, are prescribed individually rather than for the group. To date, that technology has not impacted our practice in this area.

The application of the estimated allowance is a matter of some dispute. Typically, the *PDF* is expressed as a percentage (*PDF* = 5%, 5%, 5%). The conversion from normal to standard time then would be, standard time = (normal) time + (.15)(normal) time, or Standard = (1.15)(Normal). One must assume, in this case, that *PDF* was expressed *as a percentage of normal time*. However, most allowances are estimated *as a percent-*

age of the total day, that is, a 15 percent allowance should provide for .15 × 480 minutes, or 72 minutes of nonproductive time per day. In order to apply 72 nonproductive minutes to a 480-minute day, the percentage of normal is expressed,

$$\frac{72 \text{ nonproductive minutes}}{(480-72) \text{ productive minutes}}$$

or

$$17.6\% \text{ not } 15\%,$$

that is,

$$\text{Standard} = (1.176)(\text{Normal}).$$

The recommended solution to this difficulty is to express *PDF* in *minutes per day* and use the appropriate computation to apply the allowance to normal. The alternative formulation, when *PDF* is expressed as a percentage is, $\frac{100}{100 - \%PDF}$. In this case that would be

$$100/100-15 = 1.176 \text{ or Standard} = (1.176)(\text{Normal}).$$

Your author has found that many individuals, including trained industrial engineers, have difficulty conceptualizing the difference between *PDF* as a percentage of *normal* or *productive* time and *PDF* as a percentage of *total* time. This is especially true in practice where many companies have, for years, stated PDF as a percentage of total time, but have computed the allowance as a percentage of normal time.

Improper Method

The analyst is faced with a problem if he or she observes a deviation from the prescribed method while the study is in progress. Actually, there are three types of activity elements which occur during almost every time study: (1) cyclical elements, (2) noncyclical elements, and (3) foreign elements.

Cyclical elements are part of the prescribed method and are expected to occur in every cycle. An experienced operator usually develops a rhythm and most cyclical elements are easy to time and exhibit small variance. Those elements which were circled during the study to indicate an unusual condition should be handled with caution. Truly unusual instances can be eliminated from inclusion on a select time basis. If, however, there are frequent circled elements, the analyst should make a change in the task to eliminate the cause or the times should be included.

Noncyclical elements can occur regularly or not, but in both cases they are part of the prescribed method. The frequency of occurrence of both types of elements must be recorded. This frequency or probability of occurrence is multiplied by the average observed time to give an expected time *per cycle*. The estimation of the frequency can be a source of error. Irregular noncyclical elements should be measured using work sampling.

Foreign elements are *not* part of the prescribed method. However, in the continuous observation study, the analyst must account for all of the elapsed time. As a result, when an unexpected element is introduced, the analyst notes it in the appropriate place on the data sheet, and records the end point. These activities, by definition, are not incorporated in the standard.

Another source of error due to deviation in method involves repeated actions (riveting, filing, hammering, cutting, etc.). The operator can use an excessive number of strokes while the study is underway in order to obtain a loose standard. One way to combat this problem is for the analyst to count strokes. The allowed number of strokes can be determined outside the study and the time corrected using the normal time per stroke from the study and the allowed frequency.

There are occasions in which certain activity elements can be performed internal to or during another activity element, especially during machine controlled cycles. The specified method should require the concurrent action, provided it is safe and actually feasible, since the operator will undoubtedly use that method in actual practice.

The *last resort* in correcting for improper method is to adjust the performance rating. This is an undesirable practice, at best, and should be employed only under circumstances in which there remains no alternative for the analyst. When the deviation cannot be accounted for or adjusted in any other way, the analyst may assign a performance rating factor which incorporates both pace and method.

COMPUTATIONS

The mathematics of time study is simple. The relationships which have been articulated throughout this chapter can be reduced to three basic formulas:

Normal time = average observed time × performance rating
Allowances = normal time × *PDF* factor
Standard time = normal time + allowances

The following example is a time study of a turning operation. There are seven activity elements specified in the prescribed method which appear in Figure 4-5. Activity elements **a** through **e** are cyclical elements. Element **f** is a regular, noncyclical element, and element **g** is irregular, noncyclical-estimated to occur once per ten cycles.

The results of the study are shown in Figure 4-5. Note that at the beginning of cycle 6, the operator stopped to get a drink of water, a foreign element. In cycle 8 he dropped the wrench before tightening the chuck. Also, in cycle 9, the analyst missed one of the readings. These three incidents are deviations from the usual procedure.

DATE 7 Feb 1975
TIME START 8:35 A.M.
TIME STOP 8:56 A.M.
ELAPSED TIME 21
NUMB. PIECES 11

TIME STUDY OBSERVATION

SHEET NO. 1
NO. SHEETS 1

ELEMENTS: a Load Chuck; b Advance Tool Engage Feed; c Turn 3" O.D.; d Retract Tool; e Unload Chuck; f Get 5 Shafts; g Check O.D. & Adjust Depth

NOTES	Line	a T	a R	b T	b R	c T	c R	d T	d R	e T	e R	f T	f R	g T	g R
	1	20	20	9	29	75	104	7	11	32	43				
	2	18	61	8	69	73	242	7	49	30	79				
	3	20	99	8	307	73	80	8	88	35	423			80	503
	4	19	22	8	30	75	605	6	11	32	43				
	5	21	64	9	73	74	741	7	54	32	86	60	846		
	6	20	A 80	10	90	75	1265	8	73	30	1303				
	7	22	25	9	34	73	1407	7	14	30	44				
Drop wrench	8	(28)	(72)	8	80	70	1550	7	57	35	92				
	9	19	17	-	—	-	90	6	96	34	1730				
	10	20	50	10	60	73	1833	7	40	33	73	55	1928		
	11	18	46	9	55	73	2008	8	36	29	65				
	12														
	13														
	14														
	15														
	16														

SUMMARY

	a	b	c	d	e	f	g
Total Time	197	88	734	78	352	115	8
No. of Readings	10	10	10	11	11	2	1
Av. of Readings	19.7	8.8	73.4	7.09	32.0	57.5	80
Frequency	1/1	1/1	1/1	1/1	1/1	1/5	1/11
Average Time	19.7	8.8	73.4	7.09	32.0	11.5	8.8
El. Rating Factor	110	110	135	110	110	95	135
El. Normal Time	21.67	9.68	99.09	7.80	35.20	10.93	9.83

FOREIGN ELEMENTS

SYM	R	T	DESCRIPTION
A	1160 / 846	314	Get drink
B			
C			
D			
E			
F			
G			
H			
I			

Av. Cycle Time ______
Cycle Rating Factor ______
Normal Cycle Time ______

Percent Allowances

Pers.	Fat.	Delay	Total

Std. Time Per Unit ______
Pieces Per Hour ______
Std. Hours Per 100 ______

Allowances in Minutes

Pers.	Fat.	Delay	Total
15	15	15	45

Avail. Prod. Min. Per Hr. ______
Pieces Per Hour ______
Std. Hrs. Per 100 ______

FIGURE 4-5 Time Study Observation Sheet

For activity element **a**, Load Chuck, the following calculations were performed (all times in hundredths of a minute).

Total Time = Σ observed times = 197
No. of Readings = 10 (omit Cycle 8)
Average observed time = 19.7
Normal time = $19.7 \times 1.10 = 21.67$

The same computation is repeated for each cyclical element. The only modification for noncyclical elements is the adjustment for frequency of occurrence. Notice also that activity element **c** is machine controlled. In this case, the element must be performance rated at 135 percent which is the expected incentive performance level.

The total normal time per cycle is the sum of the elemental times.

$$21.67 + 9.68 + 99.09 + 7.80 + 35.20 + 10.93 + 9.83 = 194.20 \text{ or } 1.942 \text{ min./cycle.}$$

To compute the allowance factor, we express the *PDF* time as a fraction of productive time. In this case a total of forty-five minutes *PDF* is allowed for the entire day.

$$PDF \text{ Factor} = \frac{45}{480-45} = .103$$
Allowance = (1.942)(.103) = .200 min.
Standard time = (1.942) + (.200) = 2.142 min./cycle.

Determination of the standard number of units per day is simply 480 minutes per day/2.142 minutes per cycle = 224.1 cycles per day. If the process generates one unit per cycle, the standard output would be 224 units per day.

More precise estimates of *expected* daily output can be obtained using additional knowledge about average pace. For example, if one expected an operator to average 135 percent over the day, the number of units which would be produced is computed as follows:

normal operator controlled time = .951 min
normal machine controlled time = .991 min
                                   1.942 total normal time.

At 135 percent, the *actual* operator controlled time is,

(actual time)(performance rating) = normal
(actual time) × (1.35)   = .951 min
actual time   = .704 min

The *actual* machine controlled time = .734 minutes. Thus, total time per cycle = .704 + .734 = 1.438 minutes. In one day, then, working for 435 *productive* minutes, an operator could produce 435 minutes per day/1.438 minutes per unit = 302.5 units. The simple computation of expected output, *correct only at the expected pace used to weight machine time*, would be 224.1 standard units per day × 1.35 = 302.5 units per day.

Computation of expected daily output if performance were 110 percent would be as follows:

Actual operator controlled time = $\frac{.951}{1.10}$ = .865 min.
Total cycle time = .865 + .734 = 1.599
Daily output for 435 min., 435 min. per day/1.599 min. per unit = 272 units per day.

Statistical analysis of the study uses the relationships developed in Appendix A. Element e, Unload Chuck, is selected since it is the most variable element.

$$\bar{x} = .320$$
$$\sigma = .021$$
$$\sigma/\sqrt{n} = \frac{.021}{\sqrt{11}} = .0063$$

Then, using the student t factor for $n = 11$ and 95 percent confidence, the confidence range would be

$$\bar{x} \pm t\sigma/\sqrt{n}$$
$$r = .32 \pm (2.201)(.0063) = .32 \pm .0139$$
$$r = .306 \text{ to } .334$$

If we require the range to be less than 5 percent, then,

$$\frac{.0139}{.32} = .044 \text{ or } 4.4\%$$

and the sample can be concluded to be sufficiently large. If, for some reason, ± 3 percent would be desired, then,

$$t\sigma/\sqrt{n} = (.03)(.320)$$
$$(2.201)(.021)/\sqrt{n} = .03(.320)$$
$$\sqrt{n} = \frac{(2.201)(.021)}{(.03)(.320)} = 4.815$$
$$n = 23 \text{ readings}$$

This means that a study over twice as long as the one shown would be required to provide 95 percent confidence that the true value lies with 3 percent of the elemental time of .320 minutes.

SUMMARY

Direct observation time study is the most widely used measurement technique where determination of standards for incentive pay and production control is involved. In spite of the numerous opportunities for error, the basic structure of the methodology is sound. Most of the problems which have been experienced are the result of misunderstandings or malpractice. If time studies are conducted in a *reasonable* fashion, according to accepted practice, both labor and management should benefit.

REVIEW QUESTIONS AND PROBLEMS

1. In a time study, the average observed time is 4.00 min. for element 1 and 1.00 min. for element 2. 1 was performance rated at 90 percent, 2 at 130 percent. Compute normal time.
2. Use the data from Problem 1, but assume element 2 occurs once every five cycles. Compute normal time.
3. If *PDF* allowances are 25 min. per 480-min. day, what is the standard time for Problem 1? How many units per day would be required?
4. If employees produce at 140 percent of normal, how many units would you expect to produce each day, given Problem 1?
5. A unit has a standard time of 6.50 min. Allowances are 10 percent of the total day (48 min.). What is the normal time unit?

6. If an employee would work at 125 percent for 60 min. without stopping, how many units would be produced in Problem 5?
7. If the employees in Problem 5 were hourly paid employees, how many units might you expect in an eight-hour day? (State your assumptions.)
8. A task consists of three elements. Element 2 is machine controlled. The cycle was performance rated at 90 percent and the expected earnings on incentive pace is 120 percent. Average observed time for the elements were: #1—2.00 min., #2—1.5 min., #3—1.00 min. Compute the standard time if *PDF* amount to 7 percent of the day.
9. What are the sources of error in a time study analysis?
10. Why does the analyst rate the performance of the operator?
11. What are the characteristics of activity elements in a time study?
12. Why are machine controlled elements performance rated independently?
13. How does the method which the worker employs enter into the determination of a standard time? How does one compensate for a deviation in method?

5

STANDARD DATA

Standard Data Systems are sets of normal time values for task activity elements, usually derived from direct observation time studies, which can be systematically applied to a corresponding elemental breakdown of a proposed method for a related task, in order to synthesize the time which should be allowed to perform that task.

Due to the capability of synthesis of the task time, standard data systems have a unique role in the work measurement constellation. The synthesized times can be used for cost estimation, decisions to make or buy a particular item, or determination of whether or not an activity should be undertaken. In the three preceding measurement systems the activity being measured was actually taking place before task times could be determined.

There are actually two classes of standard data, Standard Data Systems and Predetermined Time Systems. They differ primarily in the level of specificity of the activity elements and the manner in which the data are developed. Standard Data Systems (SDSs) are commonly referred to as macroscopic or simplified systems. Predetermined Time Systems (PTSs) tend to be more microscopic. PTSs are the subject of Chapter 6.

THE STANDARD DATA PROCEDURE

Step 1. State The Objectives

A key characteristic of SDSs is that they are developed in-house, using data collected within the organization which will be making the application. The objective in developing an SDS is to provide a means for synthesizing task performance times. Whenever possible the SDS approach makes use of elemental normal time values from existing time studies. This has the distinct advantage that the pace which is implied by the synthesized time standard should be inherently consistent with the concept of normal pace which prevails within the organization. The additional effort required to transform a number of individual time studies into a Standard Data System is expended in order to achieve a tailor-made means of obtaining prior knowledge of production times.

Step 2. Establish Activity Elements

Once the decision is made to develop an SDS, the analyst must begin by identifying job families. A family consists of jobs which are similar to the extent that they will have many activity elements in common. The jobs should lend themselves to generalization of the results of the study within the job family. (Mold building might be a job family in a foundry.)

Having identified the jobs which constitute the job family, the analyst must then select those specific jobs which have "good" standards from which to develop the SDS. The time studies which are selected should have withstood the test of time in actual application and meet the criteria of fairness and consistency.

Designation of activity elements which will ultimately be the building blocks of the standard data system is, in the larger sense, a planning activity. The elements which are analyzed are, of course, those found in the time studies. The more faithful the time study analyst is to the principles of elemental analysis, the easier the development of standard data will be when that need arises. At this point it might be useful for the reader to review the discussion of activity elements in Chapter 4. The additional requirement for SDS development is that the organization make a concerted effort to *standardize* elemental descriptions across time studies within the same job family. In the event that an organization anticipates development of an SDS in the future (and what organization can afford not to anticipate it?) a special effort should be made to standardize. This means that wherever possible beginning and end points of similar activity elements would be the same, regardless of which analyst performs the elemental analysis used in the original time study. In situations in which several analysts are performing time studies, analysts should cooperate on elemental breakdowns in the same way in which they should cooperate on the maintenance of the concept of normal pace.

Consider a foundry operation in which sand molds are prepared for a variety of castings. The basic preparation of a mold involves three main parts (See Figure 5-1), the cope, the drag, and the pattern. In the elemental breakdown of the operation, the first element might simply involve assembly of the cope, drag, and pattern prior to adding parting dust, sand, and packing the mold. The element would begin when the mold-maker sets the finished mold aside and end when his hands released the assembly to reach for the parting dust. A subsequent analysis would determine whether the normal time for this element varied with the size of the mold or remained constant.

An elemental analysis for a Standard Data System, then, can be characterized as a highly standardized version of an elemental analysis for a direct observation time study.

Step 3. Identify Predictor Variables

One basic task of the analyst in developing an SDS is to differentiate between those normal times which are constant across different jobs and those which vary depending on characteristics of the job, the workplace, or whatever. The specific techniques for making such a determination all

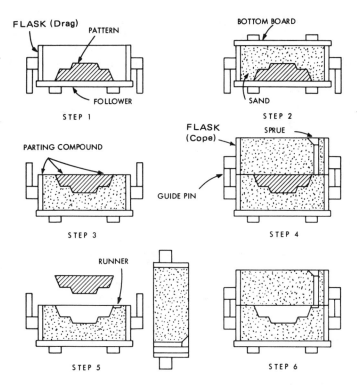

FIGURE 5-1 Principal Steps for Making a Sand Mold

require that the analyst identify those *job variables* which significantly affect the normal time required to accomplish a given activity.

The process of hypothesizing which independent or predictor variables should be investigated is actually a mixture of knowledge of the task and empirical discovery. The guideline is to always include in the study a recording of the states of those variables which your experience tells you should be included and add to them any variables which your analysis indicates will substantially reduce unexplained variance.

Ultimately the selection of predictor variables will be one of scope determination, matching the cost of SDS development and application against the accuracy and precision of the synthesized normal task times. This is discussed next.

Step 4. Determine The Scope

In an SDS, the scope of the study directly relates to the number of different time studies which are used to develop the data and the number of predictor variables which must be specified in synthesizing the new time standard.

Two of the important criteria for determination of scope which have been discussed in previous chapters have been representativeness and error. It will be helpful if, prior to addressing scope itself, we briefly review the analytical procedures of SDS development. The complete discussion is found in the following step.

One of the more convenient classifications of SDS activity elements, especially in terms of ease of application, uses four categories. The first distinction is set up versus operation. Time which occurs at the beginning or end of a run which is essentially independent of the number of cycles to be run is set up, time which varies with the number of cycles is operation time. (One clue to the distinction is the frequency of occurrence of the element, recorded on the time study.) These categories are then subdivided into fixed versus variable times. The distinction here is that elemental times are designated as fixed if one time value can be used to represent the standard time for that element in all studies analyzed and (eventually) all standards synthesized from the SDS. Variable elemental times take on different values depending on the state of one or more of the predictor variables which were hypothesized in Step 3. The analysis can be seen as a systematic approach to the classification of the activity elements and the quantification of the appropriate time relationships.

Based on this analytical framework, the determination of scope must be made in light of the following considerations.

Representativeness. The representativeness question arises directly from the classification process. The analyst must be sure that the SDS system is developed from studies which encompass a range of tasks which actually characterize the job family. For certain constant elements, it may be necessary only to analyze a few selected jobs, the objective being to confirm invariance. Caution must be exercised, however, because constant times occasionally exhibit step-like behavior. The value will remain constant over some range of conditions, but then will assume another constant value over a subsequent range. The general strategy of the analyst should be to make every effort to insure that the SDS values are representative. For variables, a larger selection may be required in order to verify the model used to describe the functional relationship between elemental time and predictor variables.

The prime concern of the analyst is, of course, that once the SDS is developed, the results can, in fact, be generalized for synthesis of new task times. The ultimate protection against inappropriate application of the SDS lies, of course, in the documentation which accompanies the SDS. It should be clear to any potential user of the system what range of task characteristics were included when the data were developed. This documentation can be referred to as specification of the boundaries of the given job family and, thus, the boundaries of the SDS.

Errors. A detailed discussion of the types of errors found in SDSs follows later in this chapter. The issue here is the relationship between errors and the scope of the analysis, specifically as it relates to the cost of the program. In all cases, we will be emphasizing the use of simple and multiple regression as the quantitative analysis tool for SDS development. Simply stated, the cost of developing and applying the SDS increases with the number of variables which must be quantified in order to use the predictive equation. The trade-off which exists in this case is the reduction in cost realized from the increased accuracy of estimation which can be achieved with more predictor values. Figure 5-2 shows this relationship and we can see the familiar U-shaped total cost curve, indicating the presence of a minimum-cost point. Analysts should be conscious of the existence of a minimum cost condition and strive to achieve at least a subjective balance between the cost of application and the accuracy of predicted task times.

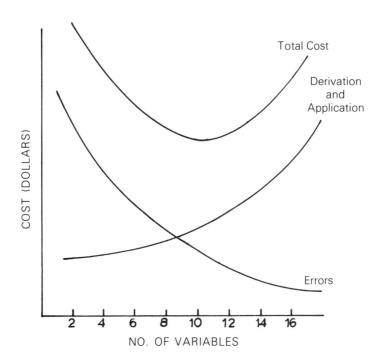

FIGURE 5-2 Cost Relationships for Standard Data

The analyst, then, is responsible (a) for specification of the boundaries which define a job family, (b) for assuring that the data which are analyzed to develop the SDS are representative of the job family task times, and (c) for incorporating those predictor variables which can economically achieve an accurate synthesis of a task time.

Step 5. Analyze The Data

The objective of the analyst is to extract from the available time study data the salient characteristics of the task which will enhance the synthesis of normal task times. In other words, the analyst seeks to identify and formalize the predictive elements in the data set.

Recall that in Step 4, the classification scheme which was described had four categories: fixed and variable set-up times and fixed and variable operating times. At this point another class should be differentiated—

operator controlled versus machine controlled activity elements. The analyses which are discussed in this chapter are applicable to operator controlled activity. Machine times which are not operator controlled constitute a separate technology. The reader is referred to the host of handbooks or other texts for this material. Often, the machining times themselves can be obtained from a manufacturer's representative for the particular equipment or the functional product or manufacturing engineer on the job in question.

Much of the analytical process can be described as an art and, as such, it is subject to the style of the analyst. In most cases, however, analysts use a combination of graphical and mathematical analysis. The plotting of various hypothesized relationships between pairs of variables permits the analyst to use those pattern-recognition capabilities which are unique to human beings as a guide to the various forms or models which might be used in the mathematical analysis.

Initially, all acceptable time studies for the designated job family should be compiled on a master data sheet or spread sheet. Table 5-1 is an example of the general format of such a spread sheet, for several elements in a job family of horizontal boring mill operations. Each entry on the sheet includes the normal elemental time, the range of the observed times, and the number of observations used to compute the average observed time. The values of the independent variables which were hypothesized in Step 3 are also recorded, as is the frequency of occurrence of the element in the cycle. Remember, since all jobs in the job family are to be recorded on the master data sheet, some jobs will not require all the activity elements provided for. This creates no problem unless there are not sufficient observations of a given element to provide the desired level of precision. If there are insufficient observations for some important elements, additional studies may be taken. Alternatively, some analysts use Predetermined Time Systems (Chapter 6) to provide supplementary times.

The next step is to make a preliminary designation of elements as variable or constant. Since constants are most easily and economically applied, the burden of proof should be on the variable. This means that either data must indicate that the times are not constant, or the analyst must suspect that the element should be a variable before subsequent analysis is indicated.

Constant elements are judged on one basic criterion: Is the average time of all elements a reasonable substitute for the actual times for each element? For a large number of elemental times, a control chart approach is suggested. Compute the average normal time for the data set. Compute the average range ($\bar{R}$) for the studies which make up the data set. Select the factor A_2 from Table 5-2 based on the smallest n recorded for any elemental time. Construct the following control limits:

$$\text{Control Limits} = \text{Average Normal Time} \pm A_2\bar{R}.$$

For Elemental 1, from Table 5-1:

$$1.500 \pm (.22)(.14) = 1.500 \pm .031$$

TABLE 5-1
SDS MASTER DATA SHEET—HORIZONTAL BORING MILL
Average Time (Min.)

Element \ Job No.	208	408	508	Frequency of Occurrence
1 — Call Crane	1.500	1.498	1.502	1/1
2 — Position Crane	.202	.201	.197	1/1
3 — Secure Chain	.430	.620	.740	1/1
4 — Adjust Chain	.079	.110	.132	1/1
5 — Hoist and Aside	1.033	.661	.988	1/1
6 — Remove Chain	.171	.202	.228	1/1
Weight of Piece	350Kg	1600Kg	2700Kg	

Range and (No. of Observations)

	208	408	508	Average Range	Smallest $\bar{R}$
1 — Call Crane	.132 (15)	.123 (16)	.165 (19)	.140	15
2 — Position Crane	.015 (15)	.015 (15)	.013 (17)	.014	15
3 — Secure Chain	.033 (15)	.071 (16)	.067 (17)	.057	15
4 — Adjust Chain	.007 (13)	.008 (16)	.008 (18)	.008	13
5 — Hoist and Aside	.113 (15)	.072 (16)	.108 (19)	.098	15
6 — Remove Chain	.020 (14)	.018 (14)	.030 (19)	.022	14

TABLE 5-2
FACTORS FOR DETERMINING 3-SIGMA LIMITS FROM $\bar{R}$

n	A_2	n	A_2
2	1.88	11	0.29
3	1.02	12	0.27
4	0.73	13	0.25
5	0.58	14	0.24
6	0.48	15	0.22
7	0.42	16	0.21
8	0.37	17	0.20
9	0.34	18	0.19
10	0.31	19	0.19
		20	0.18

Limits for $\bar{X} = \bar{\bar{X}} \pm A_2 \bar{R}$

Since none of the elemental normal times is outside that range, we can assume a stable population. (Note: If there is a large disparity in the number of observations for the various elemental times, use the weighted average rather than the mean of the normal times.) If one or more of the elemental time values had fallen outside the confidence limits, it would have been necessary for the analyst to begin a search for an assignable cause for the variation.

The quantification of variable elements takes us back to the basic approach discussed in Chapter 3, Multiple Regression Systems. Typically we seek that model which efficiently and accurately describes the relationship between one or more predictor variables and the normal task time. In MRS applications in Chapter 3, we used the coefficients of the terms in the equation as estimates of the unit task times. As such, we were interested in a model which included all the specified activity and we "forced" the model to describe the behavior of the task times. In the SDS application we revert to a more traditional application of regression analysis.

If plotting the hypothesized relationships indicates that a linear model may describe the association between dependent and independent variable, then the common strategy is to use a multiple regression analysis. Most stepwise procedures will systematically add independent variables to the analysis and select one based on some explicit criterion. For example, a forward selection technique would find the variable which produces the largest R^2. Each additional variable is introduced and an F test for statistical significance is performed. The most significant variable is then added to the regression model. The process continues until none of the remaining variables provide some specified minimum level of significance. This analytical strategy thus tests not only the simple (one-variable) model, but all other multiple variable models.

Even though the R^2 statistic is most often cited as the relevant statistic for determining the "value" of the analysis, a transform of R^2 is psychologically more meaningful to most decision makers. The transform, which we will call the Improvement Index (I) expresses the proportional reduction in variance produced by the regression model:

$$I = 1 - \sqrt{1 - R^2}$$

where R is the coefficient of simple or multiple correlation. Thus, if a regression model would, for example, produce a correlation of .67, we could say that, the model improves our certainty (reduces our uncertainty) by $1 - \sqrt{1 - (.67)^2} = .26$ or 26 percent. That is, the standard deviation about the regression line is 26 percent smaller than the standard deviation of the original data. Similarly, an R of .80 would yield a 40 percent improvement index.

For relationships which appear to be non-linear, the least-square procedure which is used to fit straight lines is also appropriate. The details of higher-order curve fitting are beyond the scope of this particular text. The interested reader is urged to the subject in any mathematical text dealing with correlation analysis and/or curve fitting.

SUMMARY OF THE PROCEDURE

Standard Data Systems provide a basis for extending the application of direct observation time study data and supplementary data to permit synthesis of standard times of related tasks within a given job family. The informational needs include: (1) a rational delineation of the characteristics of a job family, (2) a representative data base, (3) a determination of the relevant predictor variables, and (4) a quantification of the relationship between predictor variables and elemental normal times.

The reader should keep in mind that the steps outlined so far are part of the development phase of the measurement activity. In other words, the activity of the analyst culminates in a system of standard elemental times. What remains is the application phase, during which the elemental times are used to synthesize a total task time. Since the application phase of SDSs and PTSs are similar, that discussion is found in Chapter 6.

SOURCES OF ERROR

The errors which affect Standard Data Systems can be subdivided into development errors and application errors.

Development Errors

Development Errors arise as a result of the data collection and analysis activities which are necessary to build the SDS. Since the primary analytical tool for SDS development is the multiple regression technique, the section from Chapter 3 dealing with errors in MRSs should be reviewed at this time. Modeling Error, Sampling Error, and Representativeness are all present in SDSs.

In the case of modeling errors, recall that the model developed for SDS is a predictive model. It has been pointed out that we generally use stepwise procedures to select the predictor or independent variables which can be used to estimate normal times. In Chapter 3 we observed that models which use a criterion to *select* a specific set of dependent variables tend to overestimate the coefficient of correlation. This error applies to SDSs. We are computing the regression from a nonrepresentative pre-selected data set. Fortunately, such procedures do not affect the slope of the regression line. Therefore, although the "fit" may not be as good as we estimate, the relationship which is defined is not affected by the error. Of course, the matter of whether or not the model which was fit (simple linear, parabola, multiple linear, etc.) actually describes the relationship in the "real world" still must be dealt with. The test of that condition occurs at the time of application.

Another Chapter 3 observation deals with errors which result from the other type of preconditioning of data. In general, we protect ourselves from these sources of error by establishing data-selection criteria in advance. Two ways of avoiding these errors are: (1) Define the job family based on work content, not on the time values associated with the activity elements; (2) Use the test of actual application in which only standards that have survived the test of time are selected for the SDS data base. To reiterate, standards should not normally be eliminated from the data base as a result of the magnitude of the normal times.

Application Errors

Application Errors are introduced at the time the SDS is used to synthesize standard times for unmeasured work. The most serious technical error is the problem of *additive errors*.

Both experience and statistical theory tell us that when data are averaged, the errors tend to "average out." That phenomenon is quantified in the observation that the amount of variation in averages computed from samples of data is inversely proportional to the square root of the size of the sample. In other words, if there is no bias present, the averages of large samples will tend to be closer to the true value than averages of small samples.

The foregoing is *not* true of sums however. For averages, the variance is equal to or less than the variance of the elements. For a sum, the variance is equal to or greater than the elemental variances. In fact, for sums, $\sigma^2_{sum} = \sigma^2_1 + \sigma^2_2 \ldots$. Thus, with elemental normal times, we are increasing the error as we sum the elements to arrive at the total normal time. It is of no consolation whatsoever to know that the errors will tend to average out as we set a large number of standards. We must face the consequences which arise as a result of each standard we set. They affect not only our planning and cost estimation but also the earning opportunity of our incentive job employees.

Another real source of application error results from the discrete analytical approach for SDS data development. Since our analysis focuses on identification of appropriate predictor variables for activity elements, we may identify the relationship as significant for one activity element and reject it for another, not because the elemental time is not a function of the independent variable, but because the relationship is not statistically significant. In adding the elemental times for several activity elements, the cumulative error can be substantial. For example, suppose we analyze three consecutive elements from a milling operation: load chuck, mill surface, unload chuck. An analysis might be conducted to identify whether or not the elements are a function of length of cut. It is possible that the relationship would be statistically significant for only the machining operation, even though all three are, in fact, a function of the independent variable. Since we analyze each element separately the first and last would be determined to be constants, and treated as such in the standard data package.

Remedies for the nonadditivity problem are not satisfactorily addressed by the proponents of standard data systems. The rationale for this lack of apparent concern for a source of error in the system is quite pragmatic. Essentially, SDSs are used because they work. In other words, the system generally produces standard times which are not appreciably more erroneous than those set by time study. At least, if they are, the practitioners are willing to trade some lack of accuracy for the obvious advantages of prior knowledge of the standard time. Also, in the final analysis, the observation during the trial period should provide an opportunity to correct serious problems.

Your author would like to suggest an alternative, originally offered by Krick (1), which does speak to that issue. Standard Data can be developed using the multiple regression approach applied to *total cycle time* rather than to elemental times. Thus, one searches for a predictor equation to predict the standard time and the intermediate step of adding elemental times is foregone. The approach is mentioned here in hopes that it stimulates some careful exploration of this as a distinct possibility in certain situations.

COMPUTATIONS

The following example is based on data collected for Horizontal Boring Mill operations. It is an analysis of the manual unloading activity and does not include machining time. Table 5-1 is a representation of the master data sheet for the job family which was selected. The normal times are recorded for six activity elements from studies of three different part numbers. The analyst, suspecting that handling time would be affected by the weight of the units, recorded that data. Job 208 weighs 300 Kilograms, 408 weighs 1500 Kilograms, and 508 weighs 2500 Kilograms.

Initial inspection would indicate that Element 1 and Element 2 are constants. Elements 3, 4, and 6 are plotted in Figure 5-3. Each is clearly a linear function. Element 5 will require further analysis.

To compute the prediction equation for Elements 3, 4, and 6, we use the simple regression formulae:

$$(NT) = na + b \Sigma (X_1)$$

$$(NT)(X_1) = a \Sigma (X_1) + b \Sigma (X_1)^2$$

where NT = Normal Time and X_1 is the independent variable. (In this case it is the weight of the finished piece part.) For Element 3:

$$1.79 = 3a + 4650b$$
$$3140.5 = 4650a + 997 \times 10^4 b$$

yields the following:

$$NT = .391 + .000132 X_1$$

for job #208; then, with a weight of 350 kg,

$$NT = .391 + .000132(350) = .437 \text{ min.}$$

This compares to the actual time of .430 minutes, a residual error of about 1.6 percent. (The reader should compute the residual errors for other weights.) Of course, Elements 4 and 6 were analyzed in a similar fashion and their equations also appear in Figure 5-3.

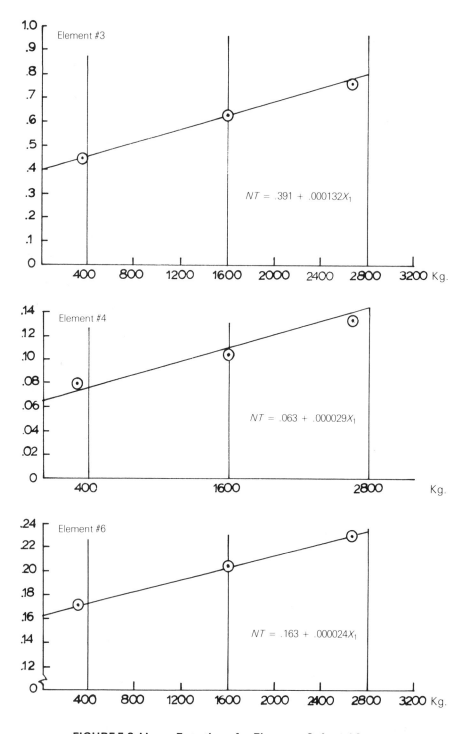

FIGURE 5-3 Linear Equations for Elements 3, 4, and 6

The normal time values for Element 5, however, were a source of some concern to the analyst. It is the element most likely to be affected by the weight of the item being handled, and preliminary analysis shows no systematic variation and large variance. It was decided that a check study should be conducted.

In recording the job conditions for the check study, the analyst discovered that the distance from the mill to the inprocess storage for finished product was substantially less than had been the case in the original study. Subsequent inquiry revealed that finished parts can be stored in any one of three locations. One was approximately ten meters, another twenty meters, and the third was about thirty meters from the mill. Three more elemental studies were conducted and the outcomes, along with the original studies are shown in Table 5-3. The analyst then performed a multiple regression analysis of the following form.

TABLE 5-3
SUPPLEMENTARY DATA (NORMAL TIMES)
Horizontal Boring Mill
Element 5. Hoist and Aside

Distance Moved \ Job No.	208	408	508
10 M	.520	.661	—
20 M	—	.903	.988
30 M	1.033	—	1.273

Normal Time × 1000 = $a + b$ (Wt. of Obj.) + c (Dis. Moved)

The output of the computer routine used to perform the multiple regression analysis is shown in Table 5-4. The resulting equation which can be used to estimate the normal time for Element 5 is:

Normal Time x 1000 =
237.157 + .097 × Wt. of Obj. + 25.466 × Dis. Moved

For job 508 at ten meters, then:

237.157 + 261.90 + 254.66 = 753.72 or .753 min.

would be the synthesized normal time for activity Element 5 on the Horizontal Boring Mill.

Once the standard data are developed for the activity elements, the analyst can commence with synthesis of standards for existing and/or contemplated jobs. Suppose, for example, a standard is sought for handling a one

TABLE 5-4
STATISTICAL ANALYSIS SYSTEM STEPWISE REGRESSION PROCEDURE FOR DEPENDENT VARIABLE *NT*

NUMBER IN MODEL	R-SQUARE	VARIABLES IN MODEL
1	0.864	DT
2	0.998	WT DT

ANALYSIS OF VARIANCE TABLE, REGRESSION COEFFICIENTS, AND STATISTICS OF FIT FOR THE ABOVE MODEL

SOURCE	DF	SUM OF SQUARES	MEAN SQUARE	F VALUE	PROB GT F	R-SQUARE	C.V.
REGRESSION	2	365264.00	182632.00	733.13	0.0002	0.998	1.76
ERROR	3	747.33	249.11				
CORRECTED TOTAL	5	366011.33					

SOURCE	DF	SEQUENTIAL SS	F VALUE	PROB GT F	PARTIAL SS	F VALUE	PROB GT F
DT	1	316406.25	1270.14	0.0001	245224.65	984.40	0.0001
WT	1	48857.75	196.12	0.0007	48857.75	196.12	0.0007

SOURCE	B VALUES	T FOR HO B=P	PROB GT ABS(T)	STD ERR B	STD B VALUES
MEAN	237.157				
DT	25.466	31.37	0.0001	0.812	0.842
WT	.097	14.00	0.0007	0.007	0.376

thousand kilogram unit for the horizontal boring mill. The particular item will be stored in Area B which is twenty meters from the mill. The synthesis is as follows:

Element	Equation	Time (min.)
1—Call Crane	Constant	1.500
2—Position Crane	Constant	.200
3—Secure Chain	.391+.000132(1000)	.523
4—Adjust Chain	.063+.000029(1000)	.092
5—Hoist and Aside	.237+.025(20)+.000097(1000)	.834
6—Remove Chain	.163+.000024(1000)	.187
	Total Time	3.336

Of course, the problem that virtually half of the total time for this activity is occupied waiting for the chain should be subjected to further analysis to determine whether or not it can be eliminated or reduced. Nevertheless, the normal time of 3.336 minutes is ready for application of appropriate allowances and inclusion into the total standard for the milling operation.

SUMMARY

Standard Data Systems extend the findings of work measurement activity in order to provide the capability to synthesize normal times. The approach is the most pragmatic of all of the measurement techniques, in that its basic rationale is that it works and the benefits of synthesis outweigh the shortcomings of questionable theoretical validity (especially nonadditivity) and lack of precision.

Depending on the homogeneity of various products, job families may be easily identified and thus provide a strong basis for development of standard data, especially in those job shop operations which are more specialized. The most significant advantage is its implicit correspondence to existing standards in the organization.

There are many possibilities for research to improve both development and application methodology in Standard Data Systems.

REVIEW QUESTIONS AND PROBLEMS

1. What is a job family? Why are job families important to standard data development?
2. What adaptations to normal time study practice are necessary for standard data development?
3. What are the significant sources of error in standard data systems?

4. Discuss this statement: "In using regression analysis for standard data, elemental time studies are not necessary."
5. Discuss the necessity of balancing the costs of derivation and application of standard data against the cost of incorrect standards, and show the cost curve(s) which illustrate the relationship.
6. In the packing department of a small company, outgoing products are batched according to customer requests. Normally from two hundred to four hundred units must be sent. Below are the observed times for a number of orders.

Time	No. of Units
26.0	198
28.5	228
31.0	245
34.5	273
37.0	310
39.5	350

Plot the time as a function of number of units. Estimate a linear equation to fit the points. Compute the same relationship using least squares. How long would it take to pack three hundred units? What might the constant term represent? Compute the improvement index.

REFERENCES

(1) Krick, E. V., *Methods Engineering*, Wiley and Sons, New York, 1966.

6

PREDETERMINED TIME SYSTEMS

Predetermined Time Systems (PTS) are a class of empirically developed elemental time values for fundamental elements of activity which can be applied to a detailed analysis of a given method for performing a task in order to synthesize the normal time required to accomplish that task.

THE PREDETERMINED TIME PROCEDURE

Step 1. State The Objective

Since the various systems which make up this body of measurement technology are invariably a product of extensive development using high-speed photography, kymography and other sophisticated instrumentation, the cost of developing and maintaining a PTS would be prohibitive for most organizations. As a result, the leading systems are virtually all marketed commercially. We generally find that the various systems are somewhat idiosyncratic, having been developed from specific industrial contexts or for certain classes of applications. Consequently, a clear statement of objectives is an important prerequisite to selection of the specific system to be used.

Typically, PTSs are chosen for two reasons: (1) Standards can be established without a direct measurement activity, and without the need for performance rating which is a source of much consternation in TSSs. (2) Standards can be synthesized with obvious benefits for costing, line balancing, etc. However, whatever the specific motivation, the objectives should be stated so that a rational decision for use of PTS rather than another work measurement system can be made.

The next decision which must be made is to select from among the competitive PTSs. Figure 6-1 is a diagram showing the origins and development of most PTSs. Of these, the three most generally used systems are Methods-Time Measurement (MTM), Work Factor, and Basic Motion Time Study (BMT). In the following presentation, MTM will be used as a prototype of Predetermined Time Systems.

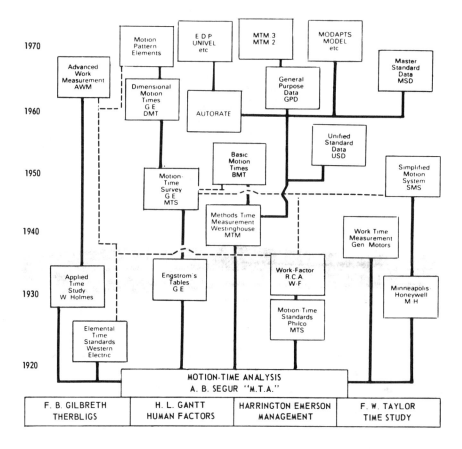

FIGURE 6-1 Family Tree of Pre-Determined Time Systems

Source: Clifford Sellie, "Family Tree of Pre-Determined Times," *History Of Motion-Times Analysis* (Chicago: Standards, International Incorporated). Reproduced with permission.

In selecting the system to be used for a specific organization, four criteria should be considered:

Cost of Installation. The application of PTS data requires a high level of training and any potential user should be certified as a practitioner by the organization which markets and maintains the data base. Cost of installation involves hiring qualified and certified analysts or having analysts trained.

Operating Cost. Most systems will provide information on application time required to set a standard using the specific system in question.

Concept of Normal. Although performance rating is not part of the PTS procedure, once a standard is set, some level of performance is associated with that standard. Different systems have different implicit performance levels. Differences between your organization's concept of normal and that of the PTS selected must be corrected through the application of a *PDF*-type of allowance.

Type of Operations. The various predetermined time systems originated in specific types of operations (electronics, metalworking). As a result one system may be better suited to your conditions than another system. A knowledge of the original data base and the industries which use a given system is helpful in making a decision in this regard.

Statement of the objective, then, includes selection of the specific PTS to be used for the conditions which prevail in the organization.

Step 2. Determine The Scope

Scope of coverage with a Predetermined Time System essentially addresses the question of coverage, that is, the total number of different jobs to be analyzed. MTM-1 is the basic system discussed in this chapter. The characteristics of jobs for which it is particularly suited are discussed later, however, in determining the scope it should be recognized that some jobs are simply not suited to such a micro-analytical approach. Common estimates are that it requires three hundred to three hundred fifty minutes of application time for each minute of performance time in using MTM-1. Consequently, MTM-1 is rarely recommended for operations with cycle times much longer than one minute, unless the volume is so large as to warrant the development cost.

Two other simplified systems, MTM-2 and MTM-3, have been developed for applications requiring less precision. MTM-2 uses only nine categories of motions which are statistical combinations of basic MTM elements and only two of these—Get and Put—are variable. Therefore, there are only thirty nine time values in the MTM-2 system and application times are in the neighborhood of one hundred fifty minutes per minute of operation time. MTM-3 has only four categories of elements and only two—Handle and Transport—have variable categories. There are ten time values in the MTM-3 system. Application times are about 50 minutes per minute of operation. MTM-2 is generally recommended for cycles of one to four minutes and MTM-3 for operation cycles greater than four minutes in duration. Many other systems, for example, Work Factor, also have macro-packages available.

Since actual use of MTM systems is not recommended unless certified and trained analysts are used, MTM-2 and -3 will not be discussed in any greater depth.

Step 3. Select The Operator

Two conditions of analysis are encountered: job in progress, standard needed; job being planned, standard needed. In either case, the analyst must observe and analyze a motion pattern. In both cases, the criteria for operator selection are identical with those discussed in Chapter 4 for Direct Observation Time Studies. Extremely low or high skill should be avoided. For TSSs, the problem lies in performance rating, for PTS the problem is the motion pattern. Low skill operators often do operations in sequence which other operators can do simultaneously. At the other extreme, highly skilled operators can often perform combinations of activities which are impossible for other operators. It takes an analyst of great experience to adjust for such non-normal methods.

The case of a planned job sometimes tempts the analyst to personally synthesize the operation. One common error is for a male analyst to use distances which may be inappropriate for jobs with predominantly female operators. However, even if the job could be synthesized without an actual operator, it is to the analyst's advantage to use one. A conscientious analyst will be painstaking in his or her exploration of alternative methods. An experienced operator can often be helpful, even for a new job, in development of the method. The fringe benefit which is obtained is the effect of having a worker participate in establishing the standard. The analyst should explain the significance of alternative methods and demonstrate the use of the tables of time values in synthesizing the standard. The educational dividends can be enormous. An operator should be used for both existing and planned jobs.

Step 4. Establish Activity Elements

It should be clear by now that use of a PTS requires the analyst to pay close attention to every detail of the motion pattern. This begins with an engineered layout of the desired work space. Principles of human factors engineering and motion economy should be applied since the time to perform the task and much of the overall productivity which can be realized will be established here. As was pointed out in Chapter 1, this text does not attempt to discuss this body of knowledge.

Once the operator and the analyst agree on an appropriate motion pattern, the analyst must record the tools, fixtures, containers, distances to and locations of anything which will be used by the operator in accomplishing the given task. Photographs are commonly used. In recording the actual motion pattern, the analyst will often film or videotape the operation after the operator has had an opportunity to practice the method.

The MTM system was originally developed for analyzing and comparing alternative work methods. Today, it is probably the most widely used of all PTSs. The system is based on eight classes of elements which are described in Figure 6-2. The analyst divides the motion pattern into elements, describes each element, and classifies each with respect to the eight classes. In addition, any conditions which might alter the times from "normal" should be recorded. The reader should review Chapter 4, Step 6 for a complete discussion of the importance of this procedure.

Step 5. Determine Elemental Times

In order to appreciate the process of assigning time values to the activity elements, it is necessary to attain a basic understanding of the MTM system. Appendix D presents the eight basic tables of MTM-1. These will be referred to frequently during the discussion. The reader should recognize that the units of this system are TMUs: .00001 hours, or .0006 minutes.

PREDETERMINED TIME SYSTEMS   87

The Maynard Research Council uses "Programmed Instruction" to teach these basic motions. The study course consists of hundreds of "Frames." Here are some frames illustrating the manual basic motions:

REACH transports the hand or fingers to a destination. When an operator transports his hand to a bolt to pick it up, his predominant purpose is to transport his hand (rather than to transport an object). Thus, the movement of the hand toward the bolt is the basic motion R E A C H . We say that the operator R E A C H E S to the bolt.

MOVE transports an object to a destination. When an operator manually transports a bolt to a totepan, his predominant purpose is to transport an object (rather than to transport his hand). The transporting of the bolt is the basic motion M O V E. We say that he M O V E D the bolt.

TURN rotates the empty or loaded hand about the long axis of the forearm. The basic motion T U R N is usually performed when turning a door knob to open a door.

APPLY PRESSURE is an application of muscular force to overcome object resistance, accompanied by little or no motion. When you press a push-button, you ordinarily do it with the basic motion A P P L Y   P R E S S U R E.

GRASP gains control of an object. If you want to use a wrench that is lying on the table, you must first G R A S P it before you can pick it up.

POSITION brings an object into an exact, predetermined relationship with another object. Locating a key in a lock is an example of P O S I T I O N.

RELEASE relinquishes control of an object. Most releases are performed merely by opening the fingers. When you are holding a hammer you R E L E A S E it by opening your fingers.

DISENGAGE separates objects when separation is followed by an involuntary movement which takes place because of the sudden ending of resistance. When an operator breaks a piece of thread in two, you observe that her hands fly apart as the resistance suddenly ends. This breaking of the thread is the basic motion D I S E N G A G E.

FIGURE 6-2 The Manual Basic Motions

Source: Reprinted with permission of The Maynard Research Council.

One of the fundamental concepts of MTM is the importance of *control* as it relates to performance. There are three types of control which affect motions:

Muscular Control. The starting and/or stopping of a motion along with any muscular action necessary to keep it on the proper path. Here muscular control pertains to the muscular action—particularly with the accuracy or precision required in that muscular action.

Visual Control. Eye action (focus and/or travel) necessary for and related to the successful completion of the motion. Often, in the lower degrees, this takes the form of orienting the operator—fixing in his mind the relation of an object (or motion destination) to himself. High visual control is characterized by exact discrimination.

Mental Control. Mental activity over and above the type needed for control of habitual motion patterns. It refers to the type of activity where simple mental decisions must be made consciously which are related directly to the performance of an individual motion. The kind of decisions referred to are not what is generally meant by planning or setting up a motion sequence, but rather the simple recurring decisions made necessary by the arrangement of or characteristics of the parts (1:49-50). The concept is particularly important in the understanding of simultaneous motions and what happens when motions requiring various levels of control are performed simultaneously. The principle here is that we describe the activity element in terms of the level of control (low, medium, high) required. These are then translated into *cases* which affect the time allocated to that element.

Before continuing with our coverage of the steps in the predetermined time procedure we should examine in detail the practical application of the concept of control as exemplified in the MTM System. This will be accomplished using two of the basic elements of motion, reach and move.

Reach. Reach is defined as the basic hand or finger motion employed when the predominant purpose is to move the hand or fingers to a destination. Three variables affect the time to perform a REACH: (1) Level of control (case), (2) Type of motion (hand in motion), (3) Distance reached (in inches).

1. Level of Control (Case)

    The *CASE A* REACH is a Reach with low control to an object or a group of objects. Generally, two conditions result in low control, objects in a fixed location (engage feed on lathe) or objects in contact with the other hand. However, if an object is fragile or sharp it may not be CASE A even though it is in the opposite hand.

    The *CASE B* REACH is a Reach with medium control to an object or a group of objects. The primary condition for CASE B is when the location of the object varies slightly from cycle to cycle; thus, vision is required. Vision is not required for termination of the reach, however, once the location is determined.

CASE C AND D REACH are Reaches with high control. From a classification standpoint, C and D are different; however, the time-distance curves coincide for the two cases. CASE C is a Reach to *an* object jumbled with other objects. The key is terminal visual control with small objects (single nut or bolt). Large objects or reach to a group of objects is not covered, nor is the selection of one item from a variety of different items which are jumbled together. CASE D involves reach to a very small object which then requires an accurate grasp. Both small object and accurate grasp are necessary for CASE D.

The *CASE E* REACH is a reach with low control to an indefinite location. Often CASE E occurs when the hand is moved out of the way or into position for some future action. CASE E reaches are rarely limiting motions.

2. Type of Motion

A *Type I* MOTION occurs when the hand is at rest at both the beginning and the end of a motion. It is the most common type.

A *Type II* MOTION occurs when the motion either begins or ends with the hand moving. Most often they involve dropping an object on the way to another object. Reach begins when the first object is dropped.

A *Type III* MOTION is quite rare, involving a motion which both begins and ends with the hand moving.

3. Distance

The DISTANCE reached is the primary variable. For FINGER reaches measure the *actual* distance moved by the tip of the finger. UNASSISTED reach involves wrist, elbow, or shoulder—the most common. The distance is the *actual* distance moved by the lowest knuckle of the index finger. (*NOT* the straight line distance.) Another class is called ASSISTED reaches. Typically assistance comes from body assistance. Without going into detail, suffice it to say that only the *actual* distance moved by the hand should be recorded.

To apply the foregoing classification, a specific notation has been developed. This permits the analyst to concentrate on the analytical task. The table look-up function can be done by a clerk or programmed for computer once the notation is specified for each activity element. A minimum of three symbols and a maximum of five symbols are used to encode a REACH. Table 6-1 provides an example of all combinations of a ten-inch reach to a fixed location. The reader is encouraged to refer to Table I in

TABLE 6-1
CODES AND TIMES FOR REACH 10"

Type of Motion	Code Symbol					Written Time (TMUs)	
	1	2	3*	4	5		
Type I		R	10	A		R10A	8.7
Type II or	m	R	10	A		mR10A	7.3
Type II		R	10	A	m	R10Am	7.3
Type III	m	R	10	A	m	mR10Am	5.9

*Use lower case f for 3/4" or less.

Appendix D to verify the time values. (Note: *Type III* MOTIONS are not included in the tables. The time is obtained by subtracting the difference between *Type I* and *Type II* from *Type III*.)

Move. Move is defined as the basic hand or finger motion employed when the predominant purpose is to use the hand to transport an object to a destination.

Four variables affect the time to perform a MOVE: (1) Level of control (case), (2) Type of motion (hand in motion), (3) Distance (in inches), and (4) Weight or resistance (in pounds).

1. Level of Control (Case)

   A move must be under the control of the operator or it is regarded process control time and measured directly. However, in moving a lever, if the motion is *not* limited by the process it is a move. Brushing chips would be a series of moves. However, if the brush were applying paint or glue requiring a specific pace, it would be regarded as process time.

   CASE A and B MOVES are moves requiring low or medium control. As was true in C and D REACH, the distinction is for classification purposes, the times are the same. CASE A is movement to the other hand or a stop. The presence of a stop results in a pattern similar to a CASE B move since the stop eliminates any positioning of the object. CASE B is movement to an approximate or indefinite location; it is the most common move encountered (lay aside piece or tool).

   CASE C MOVES are moves requiring high control. The termination is an exact location without the aid of stops or guides.

2. Type of Motion—see REACH

3. Distance—see REACH

4. Weight or Resistance (in pounds)

The weight used in analyzing MOVE is either the actual weight, if the object is carried, or the EFFECTIVE NET WEIGHT (the resistance encountered by a single hand) if resistance is the controlling factor. The move is actually accomplished in two parts, the static portion prior to any movement and the dynamic portion during which the weight is actually moving. Both the dynamic and the static components have been calibrated from two to forty pounds, and it has been determined that five pound intervals are sufficiently precise for most MTM purposes. The table values in Appendix D reflect this. Note, however, that the static component is omitted in Type II moves in which the object is in motion at the beginning or if the total weight of the object is already being held.

In determining the time to accomplish a MOVE, determine the necessary data and extract the values from the tables. The formula used is:

Time = Static Constant (if appropriate)
+ Table Value × Dynamic Factor

For an eight pound weight moved four inches against a stop,

Time = 3.9 + 6.1 × 1.11 = 10.67 TMU.

The notation for MOVE requires a maximum of six symbols, those for REACH plus one for weight, if the weight exceeds 2.5 pounds. Table 6-2 shows all combinations for an eight pound weight moved four inches to an approximate location.

There remains a comparable body of both theory and practice for the remaining motion elements in the MTM system. Once again, actual application should only be performed by analysts trained and certified in MTM. There are excellent, in-depth, treatments of the technology, however. The reader should refer to either *Basic Motions of MTM*, by Antis, Honeycutt, and Koch; (2) or *Engineered Work Measurement* by Karger and Bayha, (3) as principal references to the MTM system.

**TABLE 6-2
CODES AND TIMES FOR MOVING AN 8-LB WEIGHT
A DISTANCE OF 4 IN**

Type of Motion	Code Symbol						Written Time (TMUs)	
	1	2	3*	4	5	6		
Type I		M	4	B	8		M4B8	11.56
Type II or	m	M	4	B	8		mM4B8	4.77
Type II		M	4	B	8	m	M4B8m	8.67
Type III	m	M	4	B	8	m	mM4B8m	1.88

*Use lower case *f* for 3/4" or less.

Step 6. Compute Normal Time

The preceding step was dedicated to analysis of the elemental units of motion and their corresponding elemental time values. Even the simplest analysis of work will reveal that the motions are not performed as separate, discrete units. First, bimanual or two-handed motions are invariably involved in work activity. When two motion elements are performed at the same time by different body members, we refer to them as *simultaneous* motions. When two elements are concurrently performed by the same body member, they are called *combined* motions. The first principle in application of MTM for bimanual operations is the principle of Limiting Motion:

> If an operator performs more than one motion at a time, all of the motions can be performed in the time required by the one demanding the greatest amount of time.

In its most simple application, the rationale can be seen through an example. Suppose an operator performs medium control reaches of different distances with two hands. This would be encoded as follows:

LH	TMU	RH
(R12B)	14.4	R14B

The left hand reach (12.9 TMUs) is circled to show that it has been *limited out*. The limiting motion principle applies only to those motions which can, in fact, be performed simultaneously. Empirical evidence has been collected and is presented in the form of Table X, Appendix D, which will guide the analyst in that decision.

Research by Raphael and Clapper, (1) mentioned earlier, also yielded some information on Table X. Even though their studies showed high levels of overlap in motions, they state:

> While the above procedure (use of Table X) is not a complete and final answer, it does represent an extremely practical and workable solution to the problem, provided the procedure is viewed in proper perspective. Any list of combinations developed is not an inflexible rule, but rather a guide to the analyst that reaches maximum accuracy and practicability when the analyst gives due consideration to other factors involved—practice opportunity, job complexity, employee aptitude, and the like.

Reference to Table X shows two accommodations to the comments of Raphael and Clapper. The table is coded to indicate that certain elements can be performed simultaneously, *depending on the amount of practice*. The analyst must then decide which condition prevails. The second modification depends upon whether or not the activity in question is performed within the normal range of vision. In this case, the analyst may want to alter the work place layout to achieve simultaneity which might otherwise not be achieved.

SUMMARY OF THE PROCEDURE

Synthesis of a normal time, using a Predetermined Time System, requires the following information: (1) Knowledge of the characteristics of various PTSs and of the character of the work to which the system will be applied—leading to a selection of a specific system, (2) Training and certification in the application of the selected system, (3) Detailed motion analysis of a synthesis of a proposed task, or an operator performing an existing task, (4) Specification of the states of independent variables pertinent to the selected system, (5) Knowledge of the pace which is implicit in the selected system.

Remember, PTSs synthesize *normal times*. Allowances for personal time, fatigue, and delays must also be applied. Refer to Chapter 4 to review the application of allowances.

SOURCES OF ERROR

The errors in PTS essentially originate from two basic sources: the nature of the system, and the application of the system. However, since some of the systemic errors can, in fact, be mitigated in practice, the distinction is not as clear as might be assumed at first.

Discreteness

Discreteness errors result from the desire for tabular data, to achieve easier and faster application. For example, it was pointed out in MOVE analysis that compensation for weight is discrete in five pound intervals. There is actually a regression equation which can be used to solve the continuous situation. The problem is actually one of selecting optimum levels of specificity for a given application. Knowledgeable practitioners can, and sometimes do, resort to the equations to solve for specific time values, if they deem it necessary.

Flawless Performance

Flawless Performance is invariably assumed when the motion pattern is synthesized. Empirical evidence shows that actual motion patterns vary from cycle to cycle; the PTS analysis admits only one pattern, that specified by the analyst. In addition, errors do affect the time to perform subsequent cycles and the PTSs have no direct means of compensating for this.

Unquantified Variables

Unquantified Variables are sources of error in all PTSs. A survey of different systems shows that different systems use different independent or predictor variables. This simply means that the various system developers chose a limited set of variables to use in their system. Failure to compensate for the effects of the missing variables is a source of error. Also included in this source of error is the overall effect of learning. For example, time yourself writing your name. Now time yourself writing your name using every other letter. In theory, PTS would predict the second time would be approximately one-half the first. Empirical evidence shows it to be substantially longer, not shorter. This is an extreme example of the practice phenomenon, but the source of error should not be ignored.

Implicit Pace

Implicit Pace can present the most serious error of all. As was mentioned earlier, whenever an analyst establishes a time to accomplish a given task, there is a level of effort or pace which must be exerted in order to achieve that time per cycle. Consequently, there is a concept of normal inherent in the system being applied. In fact, basic MTM data were performance rated using a specific rating system at the outset. This implicit pace may or may not correspond to concept of normal in the user organization. The only available means of correction is the application of an additional allowance along with PDF allowances to correct for any recognized difference. The amount should be arrived at through collective bargaining, and ought to be based on information gathered from several applications of the system in the specific organizational setting. In other words, the standards ought to be tried sufficiently before adjustments are undertaken.

Judgment

Supporters of PTSs of all types promote these systems because they eliminate the need for performance rating. It is true that for those applications suited to PTSs, there is no need to performance rate a worker, when setting the standard. This is advantageous. However, hyperenthusiasts sometimes claim that the need for judgment by the analyst is removed. Based on only the limited discussions of Reach, Move, and Simultaneous Motions, the reader should have some appreciation for the judgment required in PTSs. The judgment is not removed, it is simply transferred from performance rating, to other "less odious" points of application. But judgment there is and the opportunity for observer bias is always present. In fact, the possibility of working backwards from a preconceived level of output to an analysis which supports that preconception is accomplished much more unobtrusively with PTSs than with TSSs. The same precautions which were recommended in Chapter 4 to unify the concept of normal pace should be used in PTSs. Two or more analysts should occasionally observe the same operation simultaneously. They should then discuss their analysis of class of motion, degree of simultaneity, etc.

Inefficient Method

The analyst should be thoroughly familiar with the task being analyzed. This is particularly true in the analysis of existing operations. Workers often will take advantage of an inexperienced analyst and present an inflated motion pattern. This leads to excessive normal times and subsequently to disproportionate earning opportunity. There is really no adequate substitute for a knowledgeable analyst.

Inappropriate Application

Inappropriate applications stem from two sources. First, each PTS was developed in a specific context, therefore, there are certain tasks which are not suited to the system. For example, some systems should not be applied to heavy work, others do not handle intricate, manipulative work well.

The second is inappropriate length of cycle. Micro-data applied to long cycle activities will accumulate large errors. The guidelines provided by the particular PTS should be adhered to since not only the magnitude of the errors, but also the cost of application become unacceptable.

COMPUTATIONS

The following is merely an example of the application of MTM to a representative task. A description of the method in sufficient detail to permit the reader to exactly replicate the analysis which is shown is inappropriate to the objectives of this text. However, for moves, reaches, and simultaneous motions, the reader should be able to follow the logic of the analyst.

Figure 6-3 shows the back of an MTM analysis chart with task identification and a sketch of the work place. Neither the coverplate nor the tongs nor the combination of the two is sufficiently heavy to require a weight adjustment. The front of the chart, Figure 6-4, shows the coded motion analysis. On line five, the two hands move toward each other, right hand controlling.

The bottom of the chart summarizes the MTM analysis. Element 1 is 52.0 TMUs. Converstion to hours includes 15 percent *PDF* allowance. Therefore, a normal time of 52.0 TMUs = .00052 hours, and .00052 × 1.15 = .000598 hours allowed time. (Note: 15 percent here is expressed as a percentage of *normal* time, not *total* time.)

The only lack in the analysis presented in the example is the fact that experienced operators will undoubtedly perform the elements on lines three and four sometime between line twenty-three and thirty, shortening the cycle time by 23.6 TMUs. In addition, the predominant one-handed nature of the operation should motivate the analyst to seek an improved method—perhaps redesign of the die to handle two plates with ejection—allowing bimanual operation without tongs.

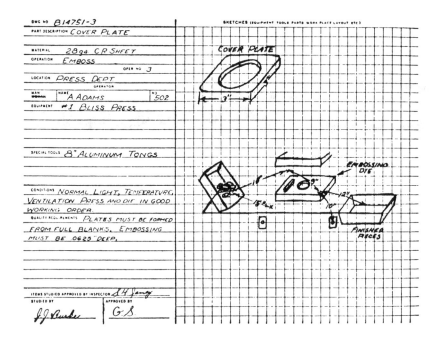

FIGURE 6-3 Back of the MTM Methods Analysis Chart

Source: Copyrighted by the MTM Association for Standards and Research. No reprint permission without written consent from the MTM Association, 9-10 Saddle River Road, Fair Lawn New Jersey 07410.

SUMMARY

Predetermined Time Systems, in the hands of a skilled analyst, are an effective tool for synthesizing task times. The skill in analysis and judgment necessary for effective application make the potential for error and the potential for abuse (by working backwards) quite high. Therefore, the requirement for ethical practice is as high for a PTS as it is in every other measurement system.

A tangible benefit is the emphasis on method which requires the analyst to concentrate on that aspect of the total work measurement and design function. Another is the precise record which is then available for training and other collateral activities.

METHODS ANALYSIS CHART

PART: Cover Plate
OPERATION: Emboss
DATE: 8-23-
ANALYST: AML
REFERENCE No: B14751-3
STUDY No: 3
SHEET No: 1 of 1 SHEETS

No.	DESCRIPTION — LEFT HAND	No.	LH	TMU	RH	No.	DESCRIPTION — RIGHT HAND
1		1.	GET PART WITH TONGS				
2							
3	Reach to part		R15C	16.3			
4			G4A	7.3			
5	Move to tongs		M18A	15.2	M12C		Move tongs to part
6				5.6	P1SE		
7				5.6	G2		Close tongs
8			RL1	2.0			
9				52.0			
10							
11		2.	PLACE PART IN DIE AND CYCLE PRESS				
12							
13	Get hand out of way		R4E	12.7	M9C		Move part to die
14				26.6	P2NSD		
15				2.0	MfB		Release part
16	Same as R/H		R2A	8.7	R10A		Reach to trip button
17					G2		Regrasp tongs
18			MfA	2.0	MfA		Press button
19				16.0	PT		Press time
20				68.0			
21							
22		3.	GET PART WITH TONGS AND LAY ASIDE IN TOTE PAN				
23	Get hand out of way		R4E	13.5	M10C		Move tongs to part
24					G2		
25				5.6	G2		Adjust tongs
26				5.6	P1SE		
27				2.0	MfA		Close tongs
28				7.5	D2E		Remove part
29				13.4	M12B		Move to tote pan
30				2.0	MiB		Release part from tongs
31				49.6			

No.	ELEMENT DESCRIPTION	ELEMENT TIME TMU	CONVERSION FACTOR .00001 LEVELED TIME	15 % ALLOWANCE	ELEMENT TIME ALLOWED Hours	OCCURRENCES PER PIECE OR CYCLE	TOTAL TIME ALLOWED Hours
1	Get part with tongs	52.0			.000598	1	.000598
2	Place part in die and cycle press	68.0			.000782	1	.000782
3	Get part and lay aside	49.6			.000570	1	.000570
						TOTAL	.00195

FIGURE 6-4 Front of the MTM Methods Analysis Chart

Source: Copyrighted by the MTM Association for Standards and Research. No reprint permission without written consent from the MTM Association, 9-10 Saddle River Road, Fair Lawn New Jersey 07410.

REVIEW QUESTIONS AND PROBLEMS

1. Why are Predetermined Time Systems usually adopted?
2. In comparing two alternative predetermined time systems, what are the issues which should be considered?

3. You have used MTM to develop a standard. Since you were not sure of your results, you took a time study. Unfortunately, the times are farther apart than you anticipated. Prepare a list of the possible items which should be investigated in order to identify the assignable cause—if one exists.
4. Estimate times for the following:
   a. R10B   d. R16A
   b. mM14B   e. mR10B
   c. M12B13   f. M4A2
5. Use the MTM tables to estimate the following—explain each step.

LH	RH
mR10B	mR10B
G4A	G4A
M10B2	M10B10

6. Explain the significance of control as it affects predetermined time data.
7. Under what conditions is a time value limited out?
8. What are the sources of error in Predetermined Time Systems?

REFERENCES

(1) Raphael, D. L., and Grant C. Clapper, *A Study of Simultaneous Motions*, MTM Association for Standards and Research, Fair Lawn, New Jersey, 1952. pp. 49-50.

(2) Antis, William, J. M. Honeycutt, Jr., and E. N. Koch, *The Basic Motions of MTM*, Fourth Edition, The Maynard Foundation, Naples, Florida, 1973.

(3) Karger, D. W., and F. H. Bayha, *Engineered Work Measurement*, Industrial Press, New York, 1957.

7
THE WORK MEASUREMENT SYSTEM

In the preceding chapters, we have studied the process of work measurement as exemplified in five measurement systems. Taking each system by itself, we focused our attention on the act of measuring. We have tried to show that the result of measuring, whether it be a standard time or simply a quantification of the worktime relationship, depends on how the process is carried out. The quality of work is a function of the measurement process in the same way that the quality of a good or a service is a function of the process which produces or delivers it. Therefore, analysts need to recognize that they have a great deal of control over the quality of work measurement and that the quality depends on the way they perform.

In this final chapter, we will step back from our preoccupation with specific measurement systems, and take a look at the more general system, the one which defines the total work measurement process.

Step 1. State The Objective

There are many reasons for measuring work. One of the first things an analyst should do is to be sure that there is a clearly articulated statement of the reasons for the application of the results of work measurement. One of the principal goals of such a listing of applications is to educate the worker. Somehow measurement becomes personalized in the mind of many workers. The measurement event is often viewed as a "him versus me" or a "them versus us" proposition. A review of some of the predominant reasons for work measurement might help to depersonalize the act.

Comparing Alternative Methods. There is more than one way to accomplish any given work task. In searching for "the one best way" to perform an operation, the analyst needs a common yardstick to use in evaluating alternatives. One such yardstick is *time per unit*.

Estimating Manpower Requirements. Once management establishes the quantity of goods or services needed in a given day or week, an estimate of time per unit can provide a basis for determining the need for overtime, extra shifts, layoffs, or personnel re-assignment to accomplish the production goals.

Determining Equipment and Space Needs. In addition to manpower requirements, the unit time values serve as a vehicle for estimating if current levels of equipment have the capacity for producing the quantities of goods needed.

Production Control and Scheduling. The important function of scheduling the various work centers and specific producing units in a manner which minimizes down time, total processing time, set-up costs, or some other performance objective, cannot take place without some estimate of production time per unit. This information is also needed for evaluating spare capacity to estimate possible delivery dates for new orders.

Cost Control. Machine costs and direct and indirect labor costs should be allocated to various units of output according to some rational process. Without knowing the amount of time which various activities consume, such a costing function becomes extremely nebulous or haphazard. Without knowledge of time per unit, management is hard pressed to decide whether specific activities are "cost effective."

Incentive Plans. No incentive plan can operate without a quantification of "a fair day's work." Of course, it is the relationship of work measurement to incentive pay systems which makes the whole process personal to the worker. It affects his or her earning potential. In the incentive pay arena, the requirements for fairness and consistency become crucial and the analyst's performance affects labor management relations and morale, in addition to the quality of management decisions with respect to the other types of applications.

The influence of the quantification of work time on the operation of an organization is pervasive. The function of a clear statement of purpose is to bring that fact to the consciousness of both labor and management.

Step 2. Assess The Environment

The work measurement environment is actually a function of the history of work measurement in a particular organization. It depends on the level of sophistication of the management and the workers. Surprisingly enough, we still find manufacturing organizations which have no history of work measurement. But going beyond the traditional arena, it is common today to find work measurement being introduced into public sector agencies. By-and-large, these agencies have no history or experience with work measurement either in management or labor. The health care industry is another sector in which analysts are being recruited to practice their work measurement skills. Experience shows that resistance to work measurement is a U-shaped function, depending on the sophistication of the organization. Figure 7-1 illustrates this phenomenon.

When an organization is new to work measurement there is little organized resistance to the analyst's activities (very often there will be no union). In an environment free of bad experiences and unwarranted fears the analyst can guide both labor and management toward a smooth relationship—unless the trust relationship has been damaged by past confrontations. At the other extreme, the new analyst coming into an organization which has an ongoing program using one or more of the measurement systems usually finds little or no organized resistance. Also, the parties have a structure for

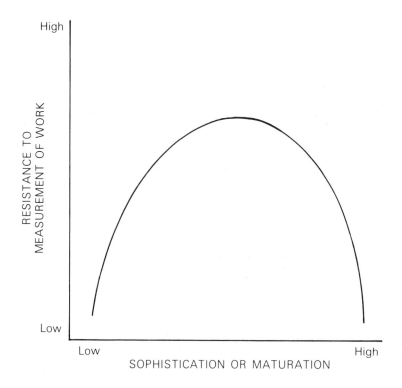

FIGURE 7-1 The Relationship Between the Sophistication of the Work Force in the Matter of Work Measurement and Their Attitude Toward It

resolving situations in which they do not agree on specific standards. The most difficulty can usually be encountered in mid-stream. Here the bad experiences often prevail, measurement is meeting active resistance, and the new analyst must play a difficult and very diplomatic role. Note that Figure 7-1 is not an inevitable progress function. A well-managed program to introduce work measurement into "virgin territory" can keep resistance at a low level throughout the maturation process.

In the early stages of implementation of a work measurement program, an approach toward broad production objectives and assessment of groups of workers rather than individuals can serve the analyst well. A *Work Sampling System* can help the analyst grasp the structure of the work environment.

Another approach to early application might involve a *Multiple Regression System*. This technique requires a certain amount of sophistication in the management due to the need for computer support and statistical expertise. However, many organizations have adopted computer technology for business and accounting purposes and remain unsophisticated in the realm of work measurement. Experience has shown that an analyst who has a firm grasp of the basic concepts of statistical inference, regression, and correlation can supply all the technical expertise necessary. The only other requisite is that management must support the measurement program and be committed to implementation. If computational support is not available in-house, service bureaus, or purchase of spare computer capacity from a bank or neighboring industry could remedy that problem.

In an organization with some measurement history, a *Time Study System* is the most frequently used work measurement system. The worker is better able to understand a standard which results from the direct observation of his or her activity. If the analyst practices in a professional manner, a work measurement system based on time study can be a powerful asset to an organization's efforts to control cost and increase productivity.

The final step in the maturation for work measurement is the introduction of the synthetic measurement methods. Where the work is appropriate, a *Predetermined Time System* or a *Standard Data System* has obvious advantages. However, the process of convincing a worker that time standards can be developed from tables is not a trivial task.

Another dimension of environment involves management's leadership style. Highly authoritarian managements are very likely to engender a hostile or resistive environment in which to practice work measurement. Here again, an analyst who understands the value of a participative approach to the work measurement process can dissipate some of the resentment, but a conflict in style between management and the analyst will always present the analyst with the problem of expending energy to combat that incongruity which might otherwise be used in more immediately productive pursuits.

The reader should not misinterpret the preceding discussion. Successful installations of and conversions to new systems, and supplementations of existing work measurement systems are always possible (and often highly desirable). In this field, as well as most others, an analyst who is sensitive to the work environment, who takes time to understand the attitudes of the management and the work force, and who takes time to inform, educate, and dispell fears wherever they surface can make a successful application of any measurement system, provided the system is compatible with the task being measured. That is the next consideration.

Step 3. Select The Appropriate System(s)

The work measurement systems which have been presented in the foregoing chapters predominate the work measurement field. An analyst who has acquired the knowledge and the skill to use the systems is equipped to handle almost any measurement problem which he or she might face in business or government today. The other requisite is proper attitude toward the task and the people involved. This has been alluded to wherever possible in our discussions, but its significance for a successful work measurement program cannot be overstressed. From the outset we have tried to present the challenge of measuring the performance of man-machine systems. Selection of a specific system for application is based on the assessment of five characteristics of a task: (1) volume of work, (2) length of cycle, (3) ease of observation of output, (4) variability of the work content, (5) physical size of the work area.

Volume of Work. Volume is the primary decision parameter, since the amount of the budget which can be feasibly expended for work measurement and process control is, in large part, dependent on volume.

Although volume and cycle time (the next parameter to be discussed) are different attributes of a job, their role in system selection is often confused. This probably stems from the fact that high volume jobs are prime candidates for mechanization and specialization, both of which tend to yield short cycle times as work is subdivided. The systems which can be best adopted to measure low volume jobs are work sampling, multiple regression, and time study. In all cases, the analyst must be sensitive to the problem of representativeness of the sample which is actually measured when low volumes of output are involved. A *Work Sampling System*, of course, attends to worker activity and thus representativeness applies more to the mix of services dictated by the demands of the job and not specifically on volume of product. A *Multiple Regression System* can be applied with an extremely small set of observations. Actually, it is more accurate to say that the simultaneous equations can be solved mathematically with small amounts of data. However, the coefficients thus derived are extremely tenuous; and until a representative sample is analyzed, it is not a good idea to even suggest that one is measuring work. Mathematical manipulation is not work measurement. With a *Time Study System*, a sample of one operator's activity is sufficient to establish a standard. Here again, we assume that the method is set, job conditions are tabulated, and an experienced worker is available to be observed. With low volume jobs, it is important to be sure that the analyst is fully prepared to measure the work when the job actually comes up for a run.

High volume jobs can support a larger fixed cost of measurement, making the use of a *Predetermined Time System* and the development of a *Standard Data System* particularly attractive. A *Time Study System* is also a legitimate contender for high volume, in fact with high volumes a case can be made for all measurement systems and other task characteristics become the critical decision variables. Interestingly enough, if there are a sufficient number of high volume jobs to justify the *PTS* or *SDS* applications, then both systems become available to synthesize the standards for the remaining low volume jobs.

Length of Cycle. The length of cycle is measured by the elapsed time required to produce one unit of output. This means that unless the job has a well-defined unit of output, cycle time is a rather meaningless concept. A short cycle-time job that is one minute or less per cycle might be high volume (run continuously) or low volume (run infrequently). Dealing only with cycle time, it is clear that short cycles are appropriate to the application of a *Predetermined Time System*. In order to apply a *Time Study System* to a short cycle job, the analyst must introduce additional technology. An elemental time study would be impossible. The technique is called *micromotion* study. The analyst uses either a motor-driven motion-picture camera or a video-tape of the operation with a clock or time display in the scene. The film is analyzed frame-by-frame or element-by-element and the results are incorporated into a time study. A *Predetermined Time System* for extremely short cycles is applied in the prescribed manner. However, if the analyst has difficulty capturing the motion pattern, a micromotion analysis can be used to study the activity elements, prior to application of the tabular data.

Long cycle jobs become very tedious for any continuous direct observation technique and a *Predetermined Time System* is extremely expensive and erratic. Thus, a *Work Sampling System* or a *Multiple Regression System* become the important approaches for the analyst to consider.

Ease of Observation of Work Output. Unless units of output can be counted, a *Multiple Regression System* cannot be employed. At the other extreme, a *Work Sampling System* can be used when outputs are delayed, or when units of output are not clearly defined (nursing services in a hospital). The remaining systems all require a clearly defined cycle, in that some combination of identifiable activities must be recorded. Therefore, there must be a cycle end point. The end point is usually associated with a unit of output, even though the output may not be immediately observable or countable.

Variability of Work Content. Some jobs can be characterized by the degree of repeatability of the content of the work from one cycle to the next. In many cases, although the time to perform the elements may vary to some extent, the sequence of motions and the sequence of the elements will remain virtually unchanged. Such jobs are usually found on mechanized or automated production lines. The systems best suited to measure such work are the *Predetermined Time Systems*, since they are deterministic and the inability to handle variability of performance is generally identified as a weakness in the systems. A *Time Study System* may also be well suited to repetitive jobs—the repetition actually makes the observer's job much easier.

Tasks with highly variable work content pose a special challenge. The usual approach is to employ a *Work Sampling System* to analyze the activities of the group involved, concentration on an assessment of inputs with concomitant recording of output. Clerical and maintenance work are typical examples of tasks commonly controlled through the sampling approach. The other measurement approach for highly variable tasks uses a *Multiple Regression System*. In these cases the analyst must be careful to indicate that large variances are inherent in the work and possibly suggest a confidence interval around the standard which is developed.

Physical Size of Work Area. When the work to be measured is confined, for the most part to a reasonably accessible area, there is no restriction on the measurement system selected, all are appropriate.

Work which involves travel, with individuals or crews scattered in remote locations or constantly moving, present particular problems for the analyst. The system which is ideally suited to this situation is a *Multiple Regression System*. Sometimes travel distances are recorded and included as an activity element in the model, or zones are mapped and the zones are included. The coefficients in these cases are hours per mile or hours per zone. Any of the other measurement systems present severe complications and the analyst ends up simply following workers around for hours with almost no structure available for analysis of the data eventually collected.

One interesting modification of a *Work Sampling System* has been incorporated in a program of work measurement in state government. The program is operated state-wide from a central location. The observation schedule has been pre-programmed and is generated daily. In addition, the group or individual to be observed is computer assigned. The analyst calls the supervisor by a Watts telephone line and the supervisor makes the observational tour using a prearranged set of activity elements. Your author is not in a position to assess the quality of the data thus generated, but it is cited here as an interesting adaptation of the *Work Sampling System* to measure work over a large physical space.

The essence of Step 3, then, is appropriate application; its goal is a truly professional response to the challenge of measuring the performance of man-machine systems.

Step 4. Measure The Work

This step simply pinpoints the point in the process when the material which has been discussed throughout this text finally comes to fruition. The analyst measures the work.

Step 5. Introduce The Results

The analyst's task is not finished when the results of his or her computations are finally available for application. The results must be introduced into the system. It matters not whether the results are being introduced for the first time in a work environment or whether they are simply a change to an existing standard. The process of introducing the results will be a disruptive influence in the work environment. Therefore, it is the responsibility of every analyst to oversee the introduction of the results of the measurement activity and to maintain that oversight until conditions "stabilize."

The first requirement has been discussed frequently in this text—a positive relationship between the analyst and the worker. By a positive relationship, we mean that the workers should have some confidence that the analyst seeks a *fair* standard. Operationally, this means that the analyst must be willing to treat the results of his or her activities as tentative. In addition, until they survive the test of actual application, all measurement values should be susceptible to re-evaluation and change. This tentative posture is often formalized in the collective bargaining agreement where incentive pay systems are involved. Typically, standards are considered to be temporary standards until a sixty to ninety day trial period is completed. During that period, either management or the union can call for a re-evaluation of the standard. After that time, unless there is a change in conditions, methods, product design, materials, or some other factor which would alter the standard by some significant amount, such as 5 percent, the value will stand.

During the trial period, the analyst should be actively involved with the operation. Questions should be answered, explanations given, and the concerns of the workers, foremen, and any other interested parties should be dealt with. It is appropriate for the analyst to have a personal stake in the standard, but that commitment should be to the standard's ultimate fairness, not to its immediate correctness. An analyst who views the introduction of a standard as an opportunity to improve labor relations has a much higher probability of long-run success.

There is a high probability that the worker's natural resistance to change will manifest itself in a cooperative "slow-down," in an effort to achieve a less demanding standard. For that reason, the trial period must be of sufficient duration to dissipate the slow-down. Employees who are used to incentive earnings will forego those earnings for only so long in order to make their point. Gradually, the "over achievers" will surface and, if the standard is fair, that fact will be demonstrated. For this reason, it is not advisable to have employees paid on average earnings for more than the first week or so of the trial period. After that, the only protection of earnings should be the principle that if the standard is judged to be too tight, earnings will be appropriately supplemented.

Step 6. Follow-up

It should be clear to all that a specific measurement result has a finite life. Work measurement is a continuous process, both in the development of new standards and in the maintenance of existing ones. A systematic approach to the audit of existing time values is the goal of every follow-up program. Somehow, however, it is a generally unrealized goal. Part of the problem may stem from the fact that groups responsible for work measurement are rarely overstaffed. Consequently, the activity is often conducted in a "fire-fighting" mode with little or no time devoted to follow-up of past work.

Two systematic processes contribute to the limited life expectancy of a particular standard. First, it is a fact of human behavior that both individual operators, collective departments, and even whole organizations exhibit "learning." The learning process is often described or modeled as a manufacturing progress function or learning curve. What is described thereby is that the time to accomplish a task will slowly decrease over extended experience with that task. As a result of the learning process, performance against a fixed standard should improve over time.

The second source of relaxation stems from the initiation of complaints by the operators who must produce to a given standard. Tight standards will be grieved and those which are tight will receive attention. There is no complementary mechanism to direct attention to loose standards. Creative employees will introduce methods changes and even modify equipment to reduce the time and effort necessary to accomplish their tasks. Unless these changes produce hazardous situations, they are rarely attended to. Perhaps the standards should not be changed under these conditions—that is matter for a book in and of itself. The point being emphasized here is that it is folly to assume that a standard which has been in effect for any extended period of time without audit actually represents the task time. If the time values which we generate when we measure work are to be considered a valid representation of the real world, follow-up is imperative.

SUMMARY

The foregoing discussion of the measurement process can be viewed as describing the operation of a general system of work measurement. In that context, then, the five measurement systems which have been treated in this text are properly considered as subsystems, all (or some) of which might be incorporated in the design of a work measurement system.

The reader is encouraged to strive to internalize the suggested systems configuration and to maintain the perspective it implies. Your author is convinced that the absence of the more general systems perspective has been a root cause (or at least a major symptom) of the low esteem in which the work measurement function is held. The lack of a systems view is manifest when organizations create positions with titles such as "time study technician." This implies that establishing time standards for work is a "time study" problem. That perception is analogous to considering plumbing malfunctions as "wrench" problems or mild headaches as "aspirin" problems.

The conclusion is not, I think, surprising. There is a real challenge to work measurement. The behavior of human beings, the synergy between people and machines, the dynamics of labor-management relations, and the limitations of the commonly used measurement systems, interact to create a milieu which defies success. However, a creative and professional approach to the challenge of work measurement as practiced by an analyst who is skilled in the technology of work measurement and sensitive to the artistic dimension of his or her activity can and will produce standards for work which are fair and consistent.

REVIEW QUESTIONS AND PROBLEMS

1. What are the six uses for the results of work measurement?
2. Explain the relationship which generally exists between the work measurement "environment" and introduction of new work measurement systems in an organization.
3. What five characteristics of a task determine the appropriateness of a particular work measurement system?
4. Select five jobs of your own choice with which you are familiar (janitor, waitress/waiter, taxi driver). Analyze the jobs according to the five selection criteria, and recommend the appropriate measurement system(s). Justify your choices.
5. Why are standards first introduced on a temporary basis?
6. "Standards become loose with age." Explain.

APPENDIX A

THEORY OF WORK SAMPLING

The mathematics of work sampling involve the understanding of the binomial distribution and the use of normal distribution as an approximation to the binomial.

Basically, events which can fall into one of only two categories and which have a probability of occurrence which does not change when the sample is taken can be represented by the binomial distribution. In its application to work sampling the two-category assumption is realized by virtue of the fact that when one observes any activity, the observation either fits into "the category of interest" or it does not. Hence, two categories.

In a work sampling study, it is the probability of occurrence (p) of a particular activity element which is our primary interest. Of secondary importance is a statement regarding the reliability of any estimate of p which is obtained from the sampling procedure.

If we make $n = 50$ observations of an activity in which a given element has a probability of occurrence, $p = .20$, then we would expect np or 10 of those observations to fall in the category of that element. In this case, np is the *expected value* or the *mean* of the binomial distribution. The standard deviation (σ) of that same binomial distribution is $\sqrt{npq}$ where $q = 1-p$. In our example

$$\sigma = \sqrt{(.50)(.20)(.80)} = 2.83.$$

However, since in work sampling our interest is actually in p not np, the distribution we employ has a mean of $np/n = p$ and a standard deviation

$$\sigma_p = \frac{\sqrt{npq}}{n} = \frac{\sqrt{pq}}{n} = \frac{\sqrt{(p)(1-p)}}{\sqrt{n}}.$$

You will recognize that it is this formula for σ_p which is used throughout your text.

One additional point should be made in this brief review of the theoretical basis for work sampling. For ease of computation the standard practice is to use the probabilities of the Normal Distribution to approximate those of the Binomial Distribution. For most situations, the Normal Approximation is acceptable if n is approximately fifty and if np is at least five.

A second occasion for statistical determination of sample size arises in the analysis of data collected for Time Study Systems. In this instance, the repeated observations of times for the activity elements constitute a sample. The issue, as it was with work sampling, is the variability in elemental time values and the precision of the mean as an estimate of the elemental time.

Hence, we have the following:

$$\text{average time}, \bar{x} = \frac{\Sigma x}{n}$$

$$\text{the standard deviation}, \sigma = \sqrt{\frac{\Sigma (x-\bar{x})^2}{n-1}}$$

$$\text{and the standard error of the mean}, \sigma_{\bar{x}} = \frac{\sigma}{\sqrt{n}}$$

where x's are individual observations and n is the number of observations.

For $n > 30$, the values of the confidence factor, k, which are based on the normal distribution are appropriate (i.e., low confidence, 1.64; moderate, 1.96; and high, 3.00). For < 30, the student's t distribution is more appropriate. For between 20 and 30 observations the values of 1.71, 2.06, and 3.65 (based on $n=25$) should suffice. Sample sizes of less than 20 observations are not recommended.

APPENDIX B

CORRELATION

If we visualize a data set and an associated regression equation, $y = f(x)$, we can express the *variation* of the dependent variable as the sum of the variation explained by the regression and the variation about the regression line (a.k.a. unexplained variation). Furthermore, we can compute two relevant *variances*: σ_y^2 is the variance computed when x is not known, $\sigma_{y \cdot x}^2$ is the variance which remains or is unexplained when x is known. Then, an expression for the proportion of the variance *not explained* by the regression would be $\sigma_{y \cdot x}^2 / \sigma_y^2$ and the corresponding expression for the proportion of variance *explained* by the regression equation would be $1 - \sigma_{y \cdot x}^2 / \sigma_y^2$. This term is sometimes referred to as the *coefficient of determination*.

The expression of the degree to which there is a functional relationship between two variables, y and x is the *coefficient of correlation* (r); which is simply the square root of the coefficient of determination:

$$r = \sqrt{1 - \frac{\sigma_{y \cdot x}^2}{\sigma_y^2}}$$

values of r fall between $+1$ and -1. The $+$ and $-$ values are indicative of the slope of a linear function and the closer the value of r is to one, the closer the data are to the regression line. The coefficient of multiple correlation (R) expresses the same quality of multivariate data that r expresses for pairs of values:

$$R_{y \cdot lmn} = \sqrt{1 - \frac{\sigma_{y \cdot lmn}^2}{\sigma_y^2}}$$

Note: Corrections are required in the foregoing equations when number of observations is small or the number of variables is large. See a standard statistics text for details.

APPENDIX C

REGRESSION EQUATIONS

The fitting of a curve to a number of data points in order to quantify the relationship between a dependent and an independent variable is most often accomplished using a least squares criterion and the techniques of simple regression.

The simultaneous equations which have been derived for fitting a straight line to a single independent variable are:

$$\sum_{i=1}^{n} y_i = b_0 n + b_1 \sum_{i=1}^{n} x_i$$

$$\sum_{i=1}^{n} x_i y_i = 1 \sum_{i=1}^{n} x_1 + b_1 \sum_{i=1}^{n} x_i^2$$

where y_i = the value of the dependent variable, data point i.
x_i = the value of the independent variable, data point i.
b_0 = the y intercept.
b_1 = the slope.

The same derivation which leads to the above equations can also be used to develop equations for fitting higher order polynomials or single dependent variables with multiple independent variables. The multiple linear regression equations can be generalized to:

$$\sum y^2 = nb_0 + b_1 \sum x_1 + b_2 \sum x_2 + b_3 \sum x_3 ... b_j \sum xy$$
$$\sum x_1 y = b_0 \sum x_1 + b_1 \sum x_1^2 + b_2 \sum x_1 x_2 + b_3 \sum x_1 x_3 ... b_j \sum x_1 x_j$$
$$\sum x_2 y = b_0 \sum x_2 + b_1 \sum x_1 x_2 + b_2 \sum x_2^2 + b_3 \sum x_2 x_3 ... b_j \sum x_2 x_j$$

$$\vdots$$

$$\sum x_j y = b_0 \sum x_j + b_1 \sum x_1 x_j + b_2 \sum x_2 x_j + b_3 \sum x_3 x_j ... b_j \sum x_j^2$$

for j independent variables and all summations are for n data points.

APPENDIX D

MTM TABLES

Do not attempt to use this chart or apply Methods-Time Measurement in any way unless you understand the proper application of the data. This statement is included as a word of caution to prevent difficulties resulting from mis-application of the data.

1 TMU	= .00001	hour
	= .0006	minute
	= .036	seconds

1 hour	= 100,000.0 TMU
1 minute	= 1,666.7 TMU
1 second	= 27.8 TMU

EFFECTIVE NET WEIGHT			
Effective Net Weight (ENW)	No. of Hands	Spatial	Sliding
	1	W	$W \times F_c$
	2	W/2	$W/2 \times F_c$
W = Weight in pounds F_c = Coefficient of Friction			

Source: Copyrighted by the MTM Association for Standards and Research. No reprint permission without written consent from the MTM Association, 9-10 Saddle River Road, Fair Lawn New Jersey 07410.

TABLE I – REACH – R

Distance Moved Inches	Time TMU				Hand In Motion		CASE AND DESCRIPTION
	A	B	C or D	E	A	B	
3/4 or less	2.0	2.0	2.0	2.0	1.6	1.6	A Reach to object in fixed location, or to object in other hand or on which other hand rests.
1	2.5	2.5	3.6	2.4	2.3	2.3	
2	4.0	4.0	5.9	3.8	3.5	2.7	
3	5.3	5.3	7.3	5.3	4.5	3.6	B Reach to single object in location which may vary slightly from cycle to cycle.
4	6.1	6.4	8.4	6.8	4.9	4.3	
5	6.5	7.8	9.4	7.4	5.3	5.0	
6	7.0	8.6	10.1	8.0	5.7	5.7	
7	7.4	9.3	10.8	8.7	6.1	6.5	
8	7.9	10.1	11.5	9.3	6.5	7.2	C Reach to object jumbled with other objects in a group so that search and select occur.
9	8.3	10.8	12.2	9.9	6.9	7.9	
10	8.7	11.5	12.9	10.5	7.3	8.6	
12	9.6	12.9	14.2	11.8	8.1	10.1	
14	10.5	14.4	15.6	13.0	8.9	11.5	D Reach to a very small object or where accurate grasp is required.
16	11.4	15.8	17.0	14.2	9.7	12.9	
18	12.3	17.2	18.4	15.5	10.5	14.4	
20	13.1	18.6	19.8	16.7	11.3	15.8	
22	14.0	20.1	21.2	18.0	12.1	17.3	E Reach to indefinite location to get hand in position for body balance or next motion or out of way.
24	14.9	21.5	22.5	19.2	12.9	18.8	
26	15.8	22.9	23.9	20.4	13.7	20.2	
28	16.7	24.4	25.3	21.7	14.5	21.7	
30	17.5	25.8	26.7	22.9	15.3	23.2	
Additional	0.4	0.7	0.7	0.6			TMU per inch over 30 inches

TABLE II – MOVE – M

Distance Moved Inches	Time TMU				Wt. Allowance			CASE AND DESCRIPTION
	A	B	C	Hand In Motion B	Wt. (lb.) Up to	Dynamic Factor	Static Constant TMU	
3/4 or less	2.0	2.0	2.0	1.7				
1	2.5	2.9	3.4	2.3	2.5	1.00	0	
2	3.6	4.6	5.2	2.9				A Move object to other hand or against stop.
3	4.9	5.7	6.7	3.6	7.5	1.06	2.2	
4	6.1	6.9	8.0	4.3				
5	7.3	8.0	9.2	5.0	12.5	1.11	3.9	
6	8.1	8.9	10.3	5.7				
7	8.9	9.7	11.1	6.5	17.5	1.17	5.6	
8	9.7	10.6	11.8	7.2				
9	10.5	11.5	12.7	7.9	22.5	1.22	7.4	B Move object to approximate or indefinite location.
10	11.3	12.2	13.5	8.6				
12	12.9	13.4	15.2	10.0	27.5	1.28	9.1	
14	14.4	14.6	16.9	11.4				
16	16.0	15.8	18.7	12.8	32.5	1.33	10.8	
18	17.6	17.0	20.4	14.2				
20	19.2	18.2	22.1	15.6	37.5	1.39	12.5	
22	20.8	19.4	23.8	17.0				
24	22.4	20.6	25.5	18.4	42.5	1.44	14.3	C Move object to exact location.
26	24.0	21.8	27.3	19.8				
28	25.5	23.1	29.0	21.2	47.5	1.50	16.0	
30	27.1	24.3	30.7	22.7				
Additional	0.8	0.6	0.85		TMU per inch over 30 inches			

TABLE III A — TURN — T

Weight	Time TMU for Degrees Turned										
	30°	45°	60°	75°	90°	105°	120°	135°	150°	165°	180°
Small — 0 to 2 Pounds	2.8	3.5	4.1	4.8	5.4	6.1	6.8	7.4	8.1	8.7	9.4
Medium — 2.1 to 10 Pounds	4.4	5.5	6.5	7.5	8.5	9.6	10.6	11.6	12.7	13.7	14.8
Large — 10.1 to 35 Pounds	8.4	10.5	12.3	14.4	16.2	18.3	20.4	22.2	24.3	26.1	28.2

TABLE III B — APPLY PRESSURE — AP

FULL CYCLE			COMPONENTS		
SYMBOL	TMU	DESCRIPTION	SYMBOL	TMU	DESCRIPTION
APA	10.6	AF + DM + RLF	AF	3.4	Apply Force
			DM	4.2	Dwell, Minimum
APB	16.2	APA + G2	RLF	3.0	Release Force

TABLE IV — GRASP — G

TYPE OF GRASP	Case	Time TMU	DESCRIPTION	
PICK-UP	1A	2.0	Any size object by itself, easily grasped	
	1B	3.5	Object very small or lying close against a flat surface	
	1C1	7.3	Diameter larger than 1/2"	Interference with Grasp on bottom and one side of nearly cylindrical object.
	1C2	8.7	Diameter 1/4" to 1/2"	
	1C3	10.8	Diameter less than 1/4"	
REGRASP	2	5.6	Change grasp without relinquishing control	
TRANSFER	3	5.6	Control transferred from one hand to the other.	
SELECT	4A	7.3	Larger than 1" x 1" x 1"	Object jumbled with other objects so that search and select occur.
	4B	9.1	1/4" x 1/4" x 1/8" to 1" x 1" x 1"	
	4C	12.9	Smaller than 1/4" x 1/4" x 1/8"	
CONTACT	5	0	Contact, Sliding, or Hook Grasp.	

TABLE V – POSITION* – P

CLASS OF FIT		Symmetry	Easy To Handle	Difficult To Handle
1–Loose	No pressure required	S	5.6	11.2
		SS	9.1	14.7
		NS	10.4	16.0
2–Close	Light pressure required	S	16.2	21.8
		SS	19.7	25.3
		NS	21.0	26.6
3–Exact	Heavy pressure required.	S	43.0	48.6
		SS	46.5	52.1
		NS	47.8	53.4
SUPPLEMENTARY RULE FOR SURFACE ALIGNMENT				
P1SE per alignment: $>1/16 \leq 1/4''$		P2SE per alignment: $\leq 1/16''$		

*Distance moved to engage—1" or less.

TABLE VI – RELEASE – RL

Case	Time TMU	DESCRIPTION
1	2.0	Normal release performed by opening fingers as independent motion.
2	0	Contact Release

TABLE VII – DISENGAGE – D

CLASS OF FIT	HEIGHT OF RECOIL	EASY TO HANDLE	DIFFICULT TO HANDLE
1–LOOSE–Very slight effort, blends with subsequent move.	Up to 1"	4.0	5.7
2–CLOSE–Normal effort, slight recoil.	Over 1" to 5"	7.5	11.8
3–TIGHT–Considerable effort, hand recoils markedly.	Over 5" to 12"	22.9	34.7
SUPPLEMENTARY			
CLASS OF FIT	CARE IN HANDLING	BINDING	
1– LOOSE	Allow Class 2	—	
2– CLOSE	Allow Class 3	One G2 per Bind	
3– TIGHT	Change Method	One APB per Bind	

TABLE VIII – EYE TRAVEL AND EYE FOCUS – ET AND EF

Eye Travel Time = $15.2 \times \frac{T}{D}$ TMU, with a maximum value of 20 TMU.

where T = the distance between points from and to which the eye travels.
D = the perpendicular distance from the eye to the line of travel T.

Eye Focus Time = 7.3 TMU.

SUPPLEMENTARY INFORMATION

– Area of Normal Vision = Circle 4" in Diameter 16" from Eyes

– Reading Formula = 5.05 N Where N = The Number of Words.

TABLE IX – BODY, LEG, AND FOOT MOTIONS

TYPE		SYMBOL	TMU	DISTANCE	DESCRIPTION
LEG–FOOT MOTION		FM	8.5	To 4"	Hinged at ankle.
		FMP	19.1	To 4"	With heavy pressure.
		LM—	7.1	To 6"	Hinged at knee or hip in any direction.
			1.2	Ea. add'l inch	
HORIZONTAL MOTION	SIDE STEP		*	≤12"	Use Reach or Move time when less than 12". Complete when leading leg contacts floor.
		SS—C1	17.0	12"	
			0.6	Ea. add'l inch	
		SS—C2	34.1	12"	Lagging leg must contact floor before next motion can be made.
			1.1	Ea. add'l inch	
	TURN BODY	TBC1	18.6	—	Complete when leading leg contacts floor.
		TBC2	37.2	—	Lagging leg must contact floor before next motion can be made
	WALK	W—FT	5.3	Per Foot	Unobstructed.
		W—P	15.0	Per Pace	Unobstructed.
		W—PO	17.0	Per Pace	When obstructed or with weight.
VERTICAL MOTION		SIT	34.7	—	From standing position.
		STD	43.4	—	From sitting position.
		B,S,KOK	29.0	—	Bend, Stoop, Kneel on One Knee.
		AB,AS,AKOK	31.9	—	Arise from Bend, Stoop, Kneel on One Knee
		KBK	69.4	—	Kneel on Both Knees.
		AKBK	76.7	—	Arise from Kneel on Both Knees.

TABLE X – SIMULTANEOUS MOTIONS

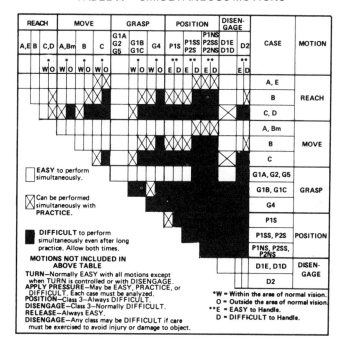

SUPPLEMENTARY MTM DATA
TABLE 1 – POSITION – P

Class of Fit and Clearance	Case of † Symmetry	Align Only	Depth of Insertion (per ¼")			
			0 >0≤1/8"	2 >1/8≤¾	4 >¾≤1¼	6 >1¼≤1¾
21 .150" – .350"	S	3.0	3.4	6.6	7.7	8.8
	SS	3.0	10.3	13.5	14.6	15.7
	NS	4.8	15.5	18.7	19.8	20.9
22 .025" – .149"	S	7.2	7.2	11.9	13.0	14.2
	SS	8.0	14.9	19.6	20.7	21.9
	NS	9.5	20.2	24.9	26.0	27.2
23* .005" – .024"	S	9.5	9.5	16.3	18.7	21.0
	SS	10.4	17.3	24.1	26.5	28.8
	NS	12.2	22.9	29.7	32.1	34.4

*BINDING—Add observed number of Apply Pressures.
DIFFICULT HANDLING—Add observed number of G2's.

†Determine symmetry by geometric properties, except use S case when object is oriented prior to preceding Move.

TABLE 1A – SECONDARY ENGAGE – E2

CLASS OF FIT	DEPTH OF INSERTION (PER 1/4")		
	2	4	6
21	3.2	4.3	5.4
22	4.7	5.8	7.0
23	6.8	9.2	11.5

TABLE 2 – CRANK (LIGHT RESISTANCE) – C

DIAMETER OF CRANKING (INCHES)	TMU (T) PER REVOLUTION	DIAMETER OF CRANKING (INCHES)	TMU (T) PER REVOLUTION
1	8.5	9	14.0
2	9.7	10	14.4
3	10.6	11	14.7
4	11.4	12	15.0
5	12.1	14	15.5
6	12.7	16	16.0
7	13.2	18	16.4
8	13.6	20	16.7

FORMULAS:
  A. CONTINUOUS CRANKING (Start at beginning and stop at end of cycle only)
$$TMU = [(N \times T) + 5.2] \cdot F + C$$
  B. INTERMITTENT CRANKING (Start at beginning and stop at end of each revolution)
$$TMU = [(T + 5.2) F + C] \cdot N$$

C = Static component TMU weight allowance constant from move table
F = Dynamic component weight allowance factor from move table
N = Number of revolutions
T = TMU per revolution (Type III Motion)
5.2 = TMU for start and stop

INDEX

Accuracy, 9
Activity elements,
   in multiple regression systems, 33-34, 36-37
   in predetermined time systems, 86-87
   in standard data systems, 68
   in time study systems, 49-50
   in work sampling systems, 16
Adaptability and compensatory beahvior, 10-11
Additive errors, 76
Allowances, 53, 59-60, 61-63
Analyzing standard data, 71
Anamorphosis, 2
Assessing work environment, 100

Basic activity elements
   in MTM, 87
   move, 90-92
   reach, 88-90
Basic motion time study, 83

Compensatory behavior and adaptability, 10-11
Computations,
   in multiple regression systems, 42-44
   in predetermined time systems, 95-97
   in standard data systems, 77-81
   in time study systems, 61-64
   in work sampling systems, 26-30
Confidence level
   in time study, 110
   in work sampling, 27, 43, 109-110
Constant elements, in standard data, 72
Continuous readings vs. snap-back, 51
Control level, in MTM, 88-89, 90
Correlation, 74, 111
Cost
   in multiple regression systems, 38
   in predetermined time systems, 84-85
   in standard data systems, 70-71

Cost *(continued)*
   in time study systems, 49
   in work sampling systems, 17-18
Cyclical elements, 60

Data errors, 75
Daywork, 6, 55-56
Delay allowances, 53-54
Development errors, 75
Discreteness errors
   in predetermined time systems, 93
   in standard data systems, 76
Distance, effect of, 89

Elemental analysis, 68
Elemental times, 86-91, 115-120
Elements of work, see activity elements
Environment of work, 100
Equipment, errors due to, 57
Equipment, time study, 57-58
Error management, 8-9
Errors, see sources of error
Ethical practice, 13
Expected performance, 47, 55-63

Fatigue allowances, 53
Follow-up, 106
Foreign elements, 61

Gomberg, William, 12
Grievances, 106

History of predetermined time systems, 83-84
Hypothesis of par, 10, 55

Implementation, 105
Improper method, 60-61
Improvement index, 74
Incentive bonuses, 12-13
Independence, 2

Information needed
  in multiple regression systems, 40
  in predetermined time systems, 93
  in standard data systems, 75
  in time study systems, 55
  in work sampling systems, 25
Interval scales, 6
Introducing results, 105

Job conditions, 50
Job families, 68
Job variables, 69
Judgment, 94

Learning, 106
Length of cycle, 103
Length of study
  in time study systems, 48, 63, 110
  in work sampling systems, 18-19, 27-28

Machine-controlled elements, 53, 59
Machining times, 72
Macro-systems of predetermined times, 85
Man-machine systems, 3, 11
Measured day work, 55-56
Measurement, 4
Measurement errors, 8-10
Measurement scales, 4-8
Measurement systems, 1, 8
Measuring work, 105
Mental control, 88
Method and performance, 11-12
Methods improvement, 12
Methods-Time-Measurement (MTM), 83
MTM tables, 115-120
Missing observations, 52
Modelling errors
  in multiple regression systems, 40
  in standard data systems, 75
Motion, classes of, 89
Move, elemental times, 90-92
Multiple regression procedure, 33-40
Multiple regression systems
  computations, 42-44
  definition of, 33
  sources of error, 40-42
  step 1, state objectives, 33-34
  step 2, define activity elements, 33-34, 36-37
  step 3, determine scope, 37-38
  step 4, establish data collection system, 38-39
  step 5, notify parties, 39
  step 6, conduct study, 40
Muscular control, 88

Nature of systems, 1-4
Negotiated allowances, 53
NLRB and Supreme Court, 13
Nominal scale, 4
Nonadditivity, 76

Noncyclical elements, 63
Nonrepresentativeness
  in multiple regression systems, 41-42
  in work sampling systems, 26
Normal pace, 47, 58, 84, 94
Normal time, 61-62
Notification of impending measurement, 13
Number of observations
  in time study systems, 48, 63, 110
  in work sampling systems, 18-19, 27-28

Objectives
  of multiple regression systems, 33-34
  of predetermined time systems, 83-84
  of standard data systems, 67
  of time study systems, 47
  of work measurement, 99
  of work sampling systems, 15
Observer bias, 26
Operator bias, 26
Operator skill, 49, 85-86
Ordinal scale, 5-6

PDF, 60
Performance and method, 11-12
Performance, flawless, 93
Performance rating, 52-53, 57-59
Personal time allowances, 53
Practice, effect of, 93
Precision, 9
Predetermined time procedure, 83-93
Predetermined time systems
  computations, 95-97
  definition of, 83
  history of, 83-84
  sources of error, 60
  step 1, state objectives, 83-84
  step 2, determine scope, 85
  step 3, select operator, 85
  step 4, define activity elements, 86-87
  step 5, determine elemental times, 86-91
  step 6, compute time, 92
Predictor variables, 68
Principle of limiting motions, 92

Randomizing observations, 20, 29
Random number table, 21
Random numbers, use of, 20-21, 29
Rating films, 58
Ratio scales, 6-7
Reach, elemental times, 88-90
Regression equations, 35-36, 42-44, 77-81, 113
Representativeness
  in multiple regression systems, 37
  in standard data systems, 70
  in time study systems, 48
  in work sampling systems, 17, 26
Resistance to change, 106
Restricting production, 56

INDEX

Sampling error
  in multiple regression systems, 42
  in work sampling systems, 25, 27, 29
Scope
  in multiple regression systems, 37-38
  in predetermined time systems, 85
  in standard data systems, 69
  in time study systems, 48
  in work sampling systems, 17-19, 28
Selecting work measurement systems, 102
Short interval scheduling, 39
Simultaneous motions, 92
Size of work area, 104
Snap-back vs. continuous readings, 51
Sources of error
  in multiple regression systems, 93-95
  in predetermined time systems, 60
  in standard data systems, 75-77
  in time study systems, 48, 57-61
  in work sampling systems, 25-26
Speed vs. accuracy, 11
Standard data procedure, 67-75
Standard data systems
  computations, 77-81
  definition of, 67
  sources of error, 75-77
  step 1, state objectives, 67
  step 2, define activity elements, 68
  step 3, identify predictor variables, 68
  step 4, determine scope, 69
  step 5, analyze data, 71-74
Standard data systems vs. predetermined time systems, 67
Standard elements, 68
Standard time, 61-63
Symbols, for motion elements, 81, 91
System, 1-2
Systems analysis, 2-3
Systems evaluation, 3

Theory of work sampling, 109-110
Time study procedure, 47-55
Time study systems
  computations, 61-64
  definition of, 47

Time study systems (*continued*)
  performance rating, 52, 58
  sources of error, 57-61
  step 1, state objectives, 47
  step 2, notify appropriate persons, 48
  step 3, determine scope, 48
  step 4, select operator, 49
  step 5, establish activity elements, 49-50
  step 6, conduct study, 50
  step 7, establish allowances, 53
Training performance rates, 58
Trial period for standards, 68

Unavoidable delay allowances, 53
Unexplained time, in multiple regression systems, 37-38
Union attitudes, 12-13
Unplanned incidents, 53
Uses of work measurement, 99-100

Variability of work content, 104
Variable work elements, 74
Visual control, 88
Volume of work, 103

Wholeness, 2
Work measurement
  definition of, 12
  objectives of, 99
  selecting appropriate systems, 102
Work sampling procedure, 15-25
Work sampling systems
  computations, 28-30
  definition of, 15
  sources of error, 25-26
  step 1, state objectives, 15
  step 2, establish activity elements, 16
  step 3, determine scope, 17-19, 28
  step 4, schedule tours, 19-22, 29
  step 5, design forms, 22
  step 6, notify appropriate parties, 22
  step 7, conduct study, 23
  Supervisor as observer, 23
Work systems, 1